D0462608

THE DISCOVERY OF GLOBAL WARMING

NEW HISTORIES OF SCIENCE, TECHNOLOGY,
AND MEDICINE

SERIES EDITORS
Margaret C. Jacob and Spencer R. Weart

SPENCER R. WEART

THE DISCOVERY OF

GLOBAL

WARMING

REVISED AND EXPANDED EDITION

HARVARD UNIVERSITY PRESS

CAMBRIDGE, MASSACHUSETTS

LONDON, ENGLAND

2008

Library of Congress Cataloging-in-Publication Data
Weart, Spencer R., 1942–
The discovery of global warming / Spencer R. Weart. —
Rev. and expanded ed.
p.   cm. — (New histories of science, technology, and medicine)
Includes bibliographical references and index.
ISBN-13: 978-0-674-03189-0 (pbk. ed. : alk. paper)
1. Global warming—History.   2. Climatic changes—History.
3. Climatology—International cooperation.   I. Title.
QC981.8.G56W43   2008
551.6—dc22   2008013675

CONTENTS

One day as I was walking home after studying scientific papers on the possibility of climate change, I noticed the elegant maples lining my street, and I wondered if they were near the southern limit of their natural range. All at once, in my mind's eye I saw the maples dead—felled by global warming.

This book is a history of how scientists came to imagine such things: the history of the science of climate change. It is a hopeful book. It tells how a few people, through ingenuity and stubborn persistence, came to understand a grave problem even before any effects became manifest. And it tells how many other people, defying the old human habit of procrastinating until a situation becomes unbearable, began working out solutions. For there are indeed ways to keep global warming within tolerable bounds with a reasonable effort. It turns out that the trees on my street are red maples, a hardy species that will hold up if we choose the proper actions soon.

The future actions we might take are not my subject. This book is a history of how we came to understand our present situation. It is an epic story: the struggles of thousands of men and women over the course of a century. For some, the work required actual physical courage, a risk to life and limb in icy wastes or on the high seas. The rest needed more subtle forms of courage. They gambled decades of arduous effort on the chance of a useful discovery and staked their reputations on what they claimed to have found. Even as they stretched their minds to the limit on intellectual problems that often proved insoluble, their attention was diverted into grueling administrative struggles to win support for the enormous task. A few took the battle into the public arena, receiving more blame than praise. Yet for all the personal drama, the effort to grasp how humanity could be changing the weather was obscure for most of a century. It was taken up by a very few individuals whose names hardly anyone had heard of; even today most of the scientists labor unknown to the world. Yet their story is as important for the future of our civilization as any history of politics and war.

If an inspector tells you that he has found termites in your house, and some day your roof will fall in, you would be a fool not to act at once. The discovery of global warming was never so clear. In 1896 a lonely Swedish scientist discovered global warming—as a theoretical concept, which most other experts declared implausible. In the 1950s a few scientists in California discovered global warming—as a possibility, a risk that might perhaps come to pass in a remote future. In 2001 an extraordinary organization mobilizing thousands of scientists around the world discovered global warming—as a phenomenon that had measurably begun to affect the weather and would probably grow much worse. That was the report from the termite inspector. But it was only one item in a tall and messy stack of reports, containing so much uncertainty and confusion that many people still wonder what, if anything, they need to do.

We have hard decisions to make. Our response to the risk of global warming will affect our personal well-being, the evolution of human society, indeed all life on our planet. One aim of this book is to help the reader understand our situation by explaining how we got here. By following how scientists in the past fought their way through the uncertainties of climate change, we can judge why they speak as they do today. More, we can understand better how scientists address the many other questions in which they have a voice.

How do scientists reach reliable conclusions? Our familiar picture of discovery, learned from the old core sciences like physics or biology, shows an orderly parade of observations and ideas and experiments. We like to think it ends with an answer, a clear statement about natural processes. Such a logical sequence, with its definitive result, does not describe work in interdisciplinary fields like the study of climate change (actually, it often doesn't describe the old core sciences, either). The story of the discovery of global warming looks less like a processional march than like a scattering of groups wandering around an immense landscape. Many of the scientists involved are scarcely aware of one another's existence. Over here we find a computer master calculating the flow of glaciers, over there an experimenter rotating a dishpan of water on a turntable, and off to the side a student with a needle teasing tiny shells out of a lump of mud. This kind of science, where specialties are only partly in contact, has become widespread as scientists labor to understand increasingly complex topics.

The tangled nature of climate research reflects Nature itself. The Earth's climate system is so irreducibly complicated that we will never grasp it completely, in the way that one might grasp a law of physics. These uncertainties infect the relationship between climate science and policy making. Debates about climate change can get as confusing as arguments over the social consequences of welfare payments. To deal with this problem, climate scientists have created remarkable new policy mechanisms. Exploring such connections between science and society at large is another aim of this book.

This is a revised edition of a book published in 2003. Much has happened since in the science and the politics, so I have expanded the last chapter into two chapters. The other chapters all reflect improved scientific and historical understanding. To keep the book short I have cut some material, adding to the omissions and oversimplifications that result from covering so much in so few pages. This book is actually a condensed version of my *real* scholarly work—a Web site of some three dozen essays containing five

times as much material as this book, including extended case studies and references to about two thousand publications, updated an-
nually. To explore the discovery of global warming more fully, visit
www.aip.org/history/climate/.

This work was supported by the American Institute of Physics and
grants from the National Science Foundation's Program in Science
& Technology Studies and the Alfred P. Sloan Foundation. I am
grateful for the indispensable help provided by scientists and histo-
rians who generously gave me interviews, comments (sometimes
extensive), or access to documents, including A. Arakawa, W.
Broecker, K. Bryan, R. A. Bryson, R. Charlson, J. Eddy, P. Edwards,
T. Feldman, J. Fleagle, J. Fleming, J. Hansen, C. D. Keeling, S.
Manabe, J. Smagorinsky, and R. M. White.

CHAPTER ONE

HOW COULD CLIMATE CHANGE?

People have always talked about the weather, but in the 1930s the talk took an unusual turn. Old folks began to insist that the weather truly wasn't what it used to be. The daunting blizzards they remembered from their childhoods back in the 1890s, the freezing of lakes in early fall, all that had ended—the younger generation had it easy. The popular press began to run articles that claimed winters really had gotten milder. Meteorologists scrutinized their records and confirmed it: a warming trend was under way. Experts told science reporters that frosts were coming later, and that wheat and codfish could now be harvested in northern zones where they had not been seen for centuries. As *Time* magazine put it in 1939, "Gaffers who claim that winters were harder when they were boys are quite right . . . weather men have no doubt that the world at least for the time being is growing warmer."[1]

Nobody worried about the change. The meteorologists explained that weather patterns always did vary modestly, in cycles lasting a few decades or centuries. If the mid-twentieth century happened to be a time of warming, so much the better. A typical popular article of 1950 promised that "vast new food-producing areas will be put under cultivation." To be sure, if the warming continued, new deserts might appear. And the oceans might rise to flood coastal cities—"another deluge, such as the catastrophe recorded in the Bible."[2] All that was plainly just colorful speculation about a remote future. Many professional meteorologists doubted that there was in fact any worldwide warming trend. All they saw were normal, temporary, regional fluctuations. And if there was a warming trend, a magazine report explained that "meteorologists do not know whether the present warm trend is likely to last 20 years or 20,000 years."[3] On 10 August 1952 the *New York Times* remarked that in thirty years people might look back fondly on the mild winters of the 1950s.

One man challenged the consensus of the experts. In 1938 Guy Stewart Callendar had the audacity to stand before the Royal Meteorological Society in London and talk about climate. Callendar was out of place, for he was no professional meteorologist, not even a scientist, but an engineer who worked on steam power. He had an amateur interest in climate and had spent many hours of his spare time putting together weather statistics as a hobby. He had confirmed (more thoroughly than anyone else) that the numbers indeed showed warming in many parts of the planet. Now Callendar told the meteorologists that he knew what was responsible. It was us, human industry. Everywhere that we burned fossil fuels, we emitted millions of tons of carbon dioxide gas (CO_2), and that was changing the climate.[4]

This idea was not new, for the basic physics had been worked out during the nineteenth century. Early in that century, the French scientist Joseph Fourier had asked himself a question. It was a deceptively simple question, of a sort that physics theory was just then beginning to learn how to attack: what determines the average temperature of a planet like the Earth? When light from the Sun strikes the Earth's surface and warms it up, why doesn't the planet keep heating up until it is as hot as the Sun itself? Fourier's answer was that the heated surface emits invisible infrared radiation, which carries the heat energy away into space. But when he calculated the effect with his new theoretical tools, he got a temperature well below freezing, much colder than the actual Earth.

The difference, Fourier recognized, was due to the Earth's atmosphere, which somehow keeps some of the heat radiation in. He tried to explain this by comparing the Earth with its covering of air to a box covered with a pane of glass. The box's interior warms up when sunlight enters and the heat cannot escape. The explanation sounded plausible, and by Callendar's time a few scientists had begun to speak of a "greenhouse effect" that keeps the Earth from freezing. It is a misnomer, for real greenhouses stay warm for their own reasons (the main effect of the glass is to keep the air, heated by sun-warmed surfaces, from wafting away). As Fourier recognized, the way the atmosphere holds in heat on the entire Earth is more subtle. The atmosphere's trick is to intercept a part of the infrared radiation emitted from the surface, preventing it from escaping into space.

The correct reasoning was first explained lucidly by a British scientist, John Tyndall. Tyndall pondered how the atmosphere might control the earth's temperature, but he was stymied by the opinion, held by most scientists at the time, that all gases are transparent to infrared radiation. In 1859 he decided to check this out in his laboratory. He confirmed that the main gases in the atmosphere, oxygen and nitrogen, are indeed transparent. He was ready to quit when he thought to try coal gas. This gas, produced by heating coal and used for lighting, was piped into his laboratory. He found that for heat rays, the gas was as opaque as a plank of wood. Thus the Industrial Revolution, intruding into Tyndall's laboratory in the form of a gas jet, declared its significance for the planet's heat balance. Tyndall went on to try other gases, and he found that the gas CO_2 was likewise opaque—what we would now call a "greenhouse" gas.

A bit of CO_2 is found in the Earth's atmosphere, and although it is only a few parts in ten thousand, Tyndall saw how it could bring warming. Just as a sheet of paper will block more light than an entire pool of clear water, so the trace of CO_2 altered the balance of heat radiation through the entire atmosphere. Much of the infrared radiation rising from the surface is absorbed by CO_2 in the air. The heat energy is transferred into the air itself rather than escaping into space. Not only is the air warmed, but also some of the energy trapped in the atmosphere is radiated back to the surface and warms it. Thus, the temperature of the Earth is maintained at a higher level than it would be without the CO_2. Tyndall put it neatly: "As a dam built across a river causes a local deepening of the stream, so our atmosphere, thrown as a barrier across the terrestrial [infrared] rays, produces a local heightening of the temperature at the Earth's surface."[5]

Tyndall's interest in all this had begun in a wholly different type of science. He hoped to solve a puzzle that was exciting great controversy among the scientists of his day: the prehistoric ice age. The claims were incredible, yet the evidence was eloquent. The scraped-down rock beds, the bizarre deposits of gravel found all around northern Europe and the northern United States—these looked exactly like the effects of Alpine glaciers, only immensely larger. Amid fierce debate, scientists were coming to accept an astounding discovery. Long ago—although not so long as geological time went, for Stone Age humans had lived through it—northern regions had been buried a mile deep in continental sheets of ice. What could have caused this?

Changes in the atmosphere were one possibility, although not a promising one. Of the atmospheric gases, CO_2 was not an obvious suspect, since there is so little of it in the atmosphere. The really important "greenhouse" gas is H_2O, simple water vapor. Tyndall found that it readily blocks infrared radiation. He explained that water vapor "is a blanket more necessary to the vegetable life of England than clothing is to man. Remove for a single summer-night the aqueous vapor from the air . . . and the sun would rise upon an island held fast in the iron grip of frost."[6] So if something dried out the atmosphere, that might cause an ice age. At present, Tyndall supposed, the atmosphere's average humidity is maintained in some sort of automatic balance, in tandem with the global temperature.

The riddle of the ice age was taken up in 1896 by a scientist in Stockholm, Svante Arrhenius. Suppose, he said, the amount of CO_2 in the atmosphere were changed. For example, a spate of volcanic eruptions might spew out vast quantities of the gas. This would raise the temperature a bit, and that small increment would have an important consequence: the warmer air would hold more moisture. Because water vapor is the truly potent greenhouse gas, the additional humidity would greatly enhance the warming. Conversely, if all volcanic emissions happened to shut down, eventually the CO_2 would be absorbed into soil and ocean water. The cooling air would hold less water vapor. Perhaps the process would spiral into an ice age.

Cooling that causes less water vapor in the air that causes more cooling that causes . . . this is the kind of self-reinforcing cycle that today we call "positive feedback." The concept was both elementary and subtle—easy to grasp, but only after somebody pointed it out. In Arrhenius's day only a few insightful scientists noticed that such effects could be crucial for understanding climate. The first important example had been worked out in the 1870s by a British geologist, James Croll, as he pondered possible causes of the ice age. When snow and ice had covered a region, he noted, they would reflect most of the sunlight back into space. Bare, dark soil and trees would be warmed by the Sun, but a snowy region would tend to remain cool. Once something started an ice age, the pattern could become self-sustaining.

Such complex effects were far beyond anyone's ability to calculate at that time. The most Arrhenius could do was to estimate the immediate effects of changing the level of CO_2. But he realized that he would also have to figure into his calculations the crucial changes in water vapor as the temperature rose or fell. The numerical computations cost Arrhenius month after month of tedious pencil work. He calculated the atmospheric moisture and the radiation entering and leaving the Earth for each zone of latitude. It seems he undertook the massive task partly as an escape from melancholy: he had just been through a divorce, losing not only his wife but custody of their little boy. The countless computations could hardly have been justified scientifically. Arrhenius had to overlook many features of the real world, and the data he used for how gases absorbed radiation were far from reliable. Nevertheless, he came up with numbers that he published with some confidence. If he was far from proving how the climate *would* change if CO_2 varied, he did in truth get a rough idea of how it *could* change. He announced that cutting the amount of CO_2 in the air by half would cool the world by maybe 5°C (that is, 8° Fahrenheit). That may not seem like a lot. But thanks to feedbacks, as extra snow accumulated and reflected sunlight, it could be enough to bring on an ice age.

Were such large changes in atmospheric composition possible? For that question Arrhenius turned to a colleague, Arvid Högbom. Högbom had compiled estimates for how CO_2 cycles through natural geochemical processes—emission from volcanoes and uptake by the oceans and so forth—and he had come up with a strange new thought. It had occurred to him to calculate the amounts of CO_2 emitted by factories and other industrial sources. Surprisingly, he found that the rate at which human activities were adding the gas to the atmosphere was roughly the same as the rates at which natural processes emitted and absorbed the gas. The added gas was not much compared with the volume of CO_2 already in the atmosphere—the amount released from the burning of coal in the year 1896 would raise the level by scarcely a thousandth part. But the additions might matter if they continued long enough. Arrhenius calculated that doubling the CO_2 in the atmosphere would raise the Earth's temperature some 5° or 6°C.

The idea of humans massively perturbing the atmosphere did not trouble Arrhenius. It was not just that warming seemed a good thing in chilly Sweden. Arrhenius, like nearly everyone at the end of the nineteenth century, expected any technological change would be for the best. People believed that scientists and engineers would solve all the problems of poverty in the centuries to come. They would turn deserts into gardens! In any case, Arrhenius figured it would take a couple of thousand years to double the amount of CO_2 in the air. In his day barely a billion people populated the world, mostly ignorant peasants living like medieval serfs. It scarcely seemed reasonable to imagine that humans could change the entire planet's atmosphere, unless perhaps in some remote and fantastic future. Arrhenius had not quite discovered global warming, but only proposed a curious theoretical concept.

Even as abstract theory, there were scientific reasons to dismiss Arrhenius's idea. Most telling was a simple laboratory measurement that seemed to refute the entire principle of greenhouse warming. An experimenter sent infrared radiation through a tube filled with CO_2. The amount of radiation that got through the tube scarcely changed when he sharply cut the quantity of gas. He knew that CO_2 absorbs radiation only in specific bands of the spectrum, and he reasoned that it took a mere trace of the gas to "saturate" these absorption bands. The atmosphere was already so thoroughly opaque there that adding more gas could make little difference. Moreover, water vapor too absorbed infrared radiation in the same general region of the spectrum. The planet evidently already showed the maximum possible greenhouse effect. By 1910 most scientists thought Arrhenius's speculation was altogether wrong.

To kill any lingering doubts, other scientists pointed out a still more fundamental objection. They held that it was impossible for CO_2 to build up in the atmosphere at all. The atmosphere is only a wisp that contains little of the material on the Earth's surface, by comparison with the huge quantities locked up in minerals and in the oceans. For every molecule of CO_2 in the air, there are about fifty dissolved in seawater. If humanity added more of the gas to the air, nearly all of it would eventually wind up in the oceans.

Furthermore, scientists saw that Arrhenius had grossly oversimplified the climate system in his calculations. For example, with more water vapor held in the air as the Earth got warmer, surely the moisture would make more clouds. The clouds would reflect sunlight back into space before the energy ever reached the surface, and so the Earth should hardly warm up after all.

These objections conformed to a view of the natural world that was so widespread that most people thought of it as plain common sense. In this view, the way cloudiness rose or fell to stabilize temperature and the way the oceans maintained a fixed level of gases in the atmosphere were examples of a universal principle: the Balance of Nature. Hardly anyone imagined that human actions, so puny among the vast natural powers, could upset the balance that governed the planet as a whole. This view of Nature—suprahuman, benevolent, and inherently stable—lay deep in most human cultures. It was traditionally tied up with a religious faith in the God-given order of the universe, a flawless and imperturbable harmony. Such was the public belief, and scientists are members of the public, sharing most of the assumptions of their culture. Once scientists found plausible arguments explaining how the atmosphere and climate would remain unchanged within a human timescale—just as everyone expected—they stopped looking for possible counterarguments.

Of course, everyone knew climate could vary. From the old folks' tales of the great blizzards of their childhood to the devastating Dust Bowl drought of the 1930s, ideas about climate included a dose of catastrophe. But a catastrophe was (by definition) something transient; things revert to normal after a few years. A few scientists speculated about greater climate shifts. For example, had a waning of rainfall over centuries caused the downfall of ancient

Near Eastern civilizations? Most doubted it. And if such changes really did happen, everyone assumed they randomly struck one region or another, not the entire planet.

To be sure, everyone knew there had been vast global climate changes in the distant past. Geologists were mapping out the ice age—or rather, ice ages. For it turned out that the tremendous sheets of ice had ground halfway down America and Europe and back not once, but over and over again. Looking still further in the past, geologists found a tropical age when dinosaurs basked in regions that were now Arctic. A popular theory suggested that the dinosaurs had perished when the Earth cooled over millions of years—climate change could be serious if you waited long enough. The most recent ice age likewise had come to a gradual end, geologists reported, as the Earth returned to its present temperature over tens of thousands of years. If a new ice age was coming, it should take as long to arrive.

The rate of advance and retreat of the great ice sheets had been no faster than present-day mountain glaciers were seen to move. That fitted nicely with "the uniformitarian principle," which held that the forces that molded ice, rock, sea, and air did not vary over time, or, as some put it, nothing could change otherwise than the way things were seen to change in the present. The principle was cherished by geologists as the very foundation of their science, for how could you study anything scientifically unless the rules stayed the same? The idea had taken hold during a century of disputes. Scientists had painfully given up traditions that explained certain geological features by Noah's Flood or other abrupt supernatural interventions. The passionate debates between "uniformitarian" and "catastrophist" theories had only partly brought science into conflict with religion. Many pious scientists and rational preachers could agree that everything happened by natural processes in a world governed by a reliable, God-given order.

Ideals of consistency pervaded not only the study of climate, but also the careers of those who studied it. Through the first half of the twentieth century, climate science was a sleepy backwater. People who called themselves "climatologists" were mostly drudges who kept track of average seasonal temperatures, rainfall, and the like. Typical were the workers at the U.S. Weather Bureau, "the stuffiest outfit you've ever seen," as one of a later generation of research-oriented geophysicists put it.[7] Their job was to compile statistics on past weather, in order to advise farmers what crops to grow or tell engineers how great a flood was likely over the lifetime of a bridge. These climatologists' products were highly appreciated by their customers (such studies continue to this day). And their tedious, painstaking style of scientific work would turn out to be indispensable for studies of climate change. Yet the value of this kind of climatology to society was based on the conviction that statistics of the previous half century or so could reliably describe conditions for many decades ahead. Textbooks started out by describing the term *climate* as a set of weather data averaged over temporary ups and downs—it was stable *by definition*.

The few who went beyond statistics to attempt explanations used only the most elementary physics. The temperature and precipitation of a region were set by the amount of sunlight at that latitude, the prevailing winds, the location of ocean currents that might warm the winds or mountain ranges that might block them, and the like. As late as 1950, if you searched a university for a climatologist, you might find one in the geography department, but not in a department of atmospheric sciences or geophysics (hardly any such departments existed anyway). The field was rightly regarded, as one practitioner complained, as "the dullest branch of meteorology."[8]

Nevertheless, plenty of vigorous speculation about climate change was heard, less from professional climatologists than from outsiders like Callendar. By the time he addressed the Royal Meteorological Society, the meteorologists had already heard only too many gaudy ideas. For although the ice ages lay in the remote past, seemingly of no practical concern, they loomed as a grand intellectual challenge. It was not the vague possibility of global warming that intrigued the few people who thought about climate change, but the stupendous advance and retreat of continental ice sheets. Tyndall, Arrhenius, Callendar, and not a few others hoped to win lasting fame by solving that notorious puzzle. From time to time newspapers would amuse their readers and embarrass climatologists by writing up some halfway plausible theory announced by one or another university professor or eccentric amateur. As one writer put it, "Everyone has his own theory—and each sounds good—until the next lad comes along with his theory and knocks the others into smithereens."[9]

When announcing theories, the professional scientists were not always easy to distinguish from the amateurs. Climatology could hardly be scientific when meteorology itself was more art than science. The best attempts to use physics and mathematics to describe weather—or even simple, regular features of the planet's atmosphere like the trade winds—had gotten nowhere. Much as climatologists could try to predict a season only by looking at the record of previous years, so meteorologists could try to predict the next day's weather only by comparison with weather of the past. Sometimes this was done systematically, matching the current weather map to an atlas of old weather maps, but more often a forecaster just looked at the existing situation and drew on experience with a combination of simple calculations, rules of thumb, and personal intuition. A canny amateur with no academic credentials could predict rain as successfully as a Ph.D. meteorologist. Indeed, through the first half of the twentieth century most of the "professionals" in the U.S. Weather Bureau lacked any college degree.

Yet it is the nature of scientists never to cease trying to explain things. If there was no accepted theory—indeed, the very word "theory" brought a skeptical frown from climatologists—there was a list of forces that could plausibly shift climate. Scientists turned up candidates everywhere, from the interior of the Earth to outer space. The possibilities spanned half a dozen different sciences.

First in line was geology. The most widely accepted explanations for the ice ages looked to the Earth's interior. If some great upheaval raised mountain ranges to block prevailing winds, for example, climates would surely change. Similarly, the raising or lowering of an island chain might change the course of the Gulf Stream so that its warmth would not reach Europe. Such forces could perhaps explain the difference between a warm age of dinosaurs and an era of ice ages. Mountain building took millions of years, however, whereas the continental ice sheets had surged back and forth in mere hundreds of thousands. To explain such relatively rapid climate shifts, geologists would have to look for other forces.

One force revealed itself in 1783, when a great volcanic eruption in Iceland poured out cubic kilometers of lava, ash, and cinders. The grass died and three-quarters of the livestock starved to death, followed by one-quarter of the people. A peculiar haze dimmed the sunlight over western Europe for months. Benjamin Franklin, visiting France, noticed the unusual cold that summer, and speculated that it might have been caused by the volcanic "fog." The idea caught on. By the end of the nineteenth century, most scientists believed that volcanic eruptions might indeed affect large regions, even the entire planet. Perhaps smoky skies during spells of massive volcanic eruptions were the cause of each advance of glaciers during the ice ages.

Other scientists suggested that the answer did not lie in the geology of volcanoes, but in the oceans. The huge bulk of the oceans contains the main ingredients of climate: far more water than the tenuous atmosphere, of course, and also most of the gases, dissolved in the seawater. And just the top few meters of the oceans hold more heat energy than the entire atmosphere. Oceanographers in the nineteenth century recognized the chief feature of the Earth's surface heat circulation. They began with the fact that water hauled up from the deep is nearly freezing, everywhere in the world. (There's a story that the inspiration for these studies came from an old practice of sailors aboard ships in the tropics: they chilled bottles of wine by dunking them overboard.) This water must have sunk in Arctic regions and flowed toward the equator along the sea bottom. The idea made sense, since water would be expected to sink where Arctic winds made it colder and therefore denser.

On the other hand, the warm tropical seas rapidly evaporated moisture, which eventually came down as rain and snow farther north or south, leaving the equatorial waters more salty. When water gets more salty it gets denser, so shouldn't ocean waters sink in the tropics? Around the turn of the century a versatile American scientist, Thomas C. Chamberlin, took an interest in the question. He figured that "the battle between temperature and salinity is a close one . . . no profound change is necessary to turn the balance."[10] Perhaps in earlier geological eras, when the poles had been warmer, salty ocean waters had plunged in the tropics and come up near the poles. This reversal of the present circulation, he speculated, could have helped maintain the uniform warmth seen in the distant past.

This was a subtle sort of explanation, and few took it up. The circulation of the oceans, like much else, was pictured as a placid equilibrium, perpetually following the same track. That was what scientists observed, if only because measurements at sea were few and difficult. Nobody had seen any reason to make the long effort that would be needed to develop a technology for making precise measurements of the seas. Oceanographers traced currents by the simple expedient of throwing bottles into the ocean. Such methods could not detect a change in the pattern of currents even if the scientists had thought to look for it. Gradually the oceanographers sketched out a pattern of stable currents. As one item in a list of many features of ocean circulation, they described how cold, dense water sinks near Iceland and Greenland and flows southward in the deep. To complete the cycle, warm water from the tropics drifts slowly northward near the surface of the North Atlantic. This had little interest for oceanographers of the day, who mainly studied rapid surface currents like the Gulf Stream. It was those that counted for practical matters like navigation and fisheries.

Another idea about causes of climate change came from an entirely different direction. Since the ancient Greeks, people had wondered whether chopping down a forest or draining a marsh might change the weather in the vicinity. It seemed common sense that shifting from trees to wheat or from wetlands to dry would affect temperature and rainfall. Americans in the nineteenth century argued that settlement of the country had brought a less savage climate, and sodbusters who moved into the Great Plains boasted that "rain follows the plow."

By the end of the nineteenth century, meteorologists had accumulated enough reliable weather records to test the idea. It failed the test. Even the transformation of the entire ecosystem of eastern North America from woods to farmland had made little evident difference to the climate. Apparently, the atmosphere was indifferent to biology.[11] That seemed reasonable enough. Whatever forces could change climate were surely far mightier than the thin scum of organic matter that covered some patches of the planet's surface.

A bare handful of scientists thought otherwise. The deepest thinker was the Russian geochemist Vladimir Vernadsky. From his work mobilizing industrial production for the First World War, Vernadsky recognized that the volume of materials produced by human industry was approaching the scale of geological processes. Analyzing biochemical activities, he concluded that the oxygen, nitrogen, and CO_2 that make up the Earth's atmosphere are put there largely by living creatures. In the 1920s he published works arguing that living organisms constituted a force for reshaping the planet comparable to any physical force. Beyond this he saw a new and still greater force coming into play: intelligence. Vernadsky's visionary pronouncements about humanity as a geological force were not widely read, however, and struck most readers as nothing but romantic ramblings.

A stronger claim to explain climate came from the seemingly most unworldly of sciences, astronomy. It began with a leading eighteenth-century astronomer, William Herschel. He noted that some stars varied in brightness, and that our Sun is itself a star. Might the Sun vary its brightness, bringing cooler or warmer periods on Earth? Speculation increased in the mid-nineteenth century, following the discovery that the number of spots seen on the Sun rises and falls in a regular eleven-year cycle. It appeared that the sunspots reflected some kind of storminess on the Sun's surface— violent activity that had measurable effects on the Earth's magnetic field. Perhaps sunspots connected somehow with weather—with droughts, for example? That would raise or lower the price of grain, so some people searched for connections with the stock market. The study of sunspots might give hints about longer-term climate shifts too.

Most persistent was Charles Greeley Abbot of the Smithsonian Astrophysical Observatory. The observatory already had a program of measuring the intensity of the Sun's radiation received at the Earth, called the "solar constant." Abbot pursued the program single-mindedly, and by the early 1920s he had concluded that the solar constant was misnamed. His observations showed large variations over periods of days, which he connected with sunspots passing across the face of the Sun. Over a term of years the more active Sun seemed brighter by nearly 1 percent. As early as 1913 Abbot had announced that he could see a plain correlation between the sunspot cycle and cycles of temperature on Earth. Self-confident and combative, Abbot defended his findings against all objections, meanwhile telling the public that solar studies would bring wonderful improvements in weather prediction. Other scientists were quietly skeptical, for the variations Abbot reported teetered at the very edge of detectability.

The study of cycles was generally popular through the first half of the twentieth century. Governments had collected a lot of weather data to play with, and inevitably people found correlations between sunspot cycles and selected weather patterns. If rainfall in England didn't fit the cycle, maybe storminess in New England would. Respected scientists and enthusiastic amateurs insisted they had found patterns reliable enough to make predictions.

Sooner or later, though, every prediction failed. An example was a highly credible forecast of a dry spell in Africa during the sunspot minimum of the early 1930s. When the period turned out wet, a meteorologist later recalled, "the subject of sunspots and weather relationships fell into disrepute, especially among British meteorologists who witnessed the discomfiture of some of their most respected superiors." Even in the 1960s, he said, "for a young [climate] researcher to entertain any statement of sun-weather relationships was to brand oneself a crank."[12] Yet *something* had caused the ice ages. Long-term cycles of the Sun were as likely a possibility as any.

It seemed there was scarcely any science that could not make a claim on climate—even celestial mechanics. In the 1870s James Croll published calculations of how the gravitational pulls of the Sun, Moon, and planets subtly affect the Earth's motions. The inclination of the Earth's axis and the shape of its orbit around the Sun oscillate gently in cycles lasting tens of thousands of years. During some millennia the Northern Hemisphere would get slightly less sunlight during the winter than it would get during other times. Snow would accumulate, and Croll argued that could reflect away enough sunlight to keep the surface cold, bringing on a self-sustaining ice age. The timing of such changes could be calculated exactly using classical mechanics (at least in principle, for the mathematics was thorny).

The glacial periods did seem to follow a cyclical pattern. The advances and retreats of ancient glaciers could be detected from long mounds of gravel (moraines) that marked where the ice had halted, and in fossil shorelines of lakes in regions that were now dry. From meticulous studies of such surface features, first in Europe and then around the world, a generation of geologists constructed a sequence. They described four distinct advances and retreats, four ice ages separated by long and equable warm ages. Croll's timing did not match this sequence at all.

Nevertheless, a few enthusiasts pursued the theory. Taking the lead was a Serbian engineer, Milutin Milankovitch. He argued that it was less in winter than in summer that extra sunlight would make a big difference in determining whether snow melted or built up into a continental ice sheet. Between the two world wars he improved the tedious calculations of the varying distances and angles of the Sun's radiation. By the 1940s some climate textbooks were teaching that Milankovitch's calculations gave a plausible solution to the problem of timing the ice ages.

Supporting evidence came from *varves,* a Swedish word for layers of silt covering the bottoms of northern lakes. A thin sheet was laid down each year by the spring runoff. Scientists extracted samples of slick, gray clay from lake beds and painstakingly counted the layers. Some researchers reported finding a 21,000-year cycle of changes. That approximately matched the timing for a cyclical skewing of the Earth's axis that Milankovitch had calculated (the "precession of the equinoxes").

But Milankovitch's numbers, like Croll's, failed to match the standard sequence of the four ice ages found in every geological textbook. Worse, there was a basic physical argument against the whole theory. The variations that Milankovitch computed in the angle and intensity of incoming sunlight were slight. Most scientists thought it far-fetched to claim that a tiny shift in sunlight, too small to be noticed by the naked eye, could bury half a continent under ice. So what had caused the ice ages? That was still anybody's guess.

Thus, when Callendar stood up before the Royal Meteorological Society in 1938, he was following many others who had speculated about climate change. Pointing to measurements of CO_2 he had dug up in old and obscure publications, he argued that the level of the gas in the atmosphere had risen a bit since the early nineteenth century. The experts were dubious. They understood that nobody had been able to make reliable measurements of the slight trace of CO_2 in the atmosphere. Callendar seemed to be picking only the data that supported his case (only in retrospect can we confirm that his judgment was pretty good). To be sure, Callendar had compiled the most convincing evidence yet that global temperatures had been rising. But was there any reason to connect the rise with CO_2?

It was not a pressing issue. Callendar himself thought a warmer climate would be a good thing for humanity, helping crops grow more abundantly. In any case, he calculated we were not raising the temperature seriously, maybe 1° by the end of the twenty-second century. The meteorologists in his audience found it all intriguing but unconvincing.

So the climate debates continued, as each expert championed a personal theory about *the* cause of climate change, the single dominant force. Most scientists gave short shrift to any theory whatsoever. They set climate change aside as a puzzle too difficult for anyone to solve with the tools at hand. The idea that humans were influencing global climate by emitting CO_2 sat on the shelf with the other bric-a-brac, a theory more peculiar and unattractive than most.

CHAPTER TWO

DISCOVERING A POSSIBILITY

Charles David Keeling—Dave to his friends—loved chemistry, and he loved the outdoors. As a postdoctoral student at the California Institute of Technology in the mid-1950s, he was committed to the sterile stinks of the laboratory, but he spent all the time he could spare traveling mountains and woodland rivers. He chose research topics that would keep him in direct contact with wild nature. Monitoring the level of CO_2 in the open air would do just that. Keeling's work was one example of how geophysics research often rested on love of the true world itself. When scientists on a lonely tundra or on a ship plowing the restless seas devoted their years to research topics that many of their peers thought of minor import, part of the reason might have been that these particular scientists could not bear to spend all their lives indoors. Yet their research sometimes turned out to be more significant than even they had hoped.

The study of atmospheric CO_2 had little to recommend itself to an ambitious scientist. Aside from the remote possibility that it might play some role in climate change over thousands of years, there was only a mild curiosity about how the winds carried around stuff that crops needed to grow, including the carbon in CO_2. A group in Scandinavia had attempted a monitoring program. Their measurements of CO_2 fluctuated widely from place to place, and even from day to day, as different air masses passed through carrying pulses of gas emitted by a forest or a factory. "It seems almost hopeless," one expert confessed, "to arrive at reliable estimates of the atmospheric carbon-dioxide reservoir and its secular changes by such measurements."[1] And regardless of Keeling's personal interest, would any agency give him money to make a new attempt?

That short question has a long and interesting answer. It begins with the revolutionary changes that the Second World War and the onset of the Cold War brought to the American scientific community. Consider, as a typical case, the transformation of meteorology. Generals and admirals, knowing well how battles can turn on the weather, needed meteorologists. The U.S. military turned to institutions like the University of Chicago, where a newly created department of meteorology was one of the few places in the world where the subject was being studied with full scientific rigor.

That was thanks to Carl-Gustav Rossby. Trained in mathematical physics in Stockholm, Rossby had come to the United States in 1925 to work in the Weather Bureau. He soon left the somnolent Bureau in disgust. Outstanding not only as a theorist but also as an entrepreneur and organizer, he created the nation's first professional meteorology program at the Massachusetts Institute of Technology. In 1942 he moved on to Chicago to establish another. Rossby was a leader of a scattered group of meteorologists who were determined to make the study of climate truly scientific. In place of the traditional climatology that merely listed descriptions of the "normal" climate in each geographical region, they would derive a more complex understanding of climate from basic principles of physics. The goal was an exercise in pure mathematics, deliberately remote from the fluctuations of actual weather and the uncertainties of daily predictions.

The scientific project was delayed by the war, but the Chicago meteorology department expanded tremendously during the war years. Rossby and his colleagues trained some 1,700 military meteorologists in one-year courses. Similarly, in other institutions and other fields of geophysics, teaching and research enterprises thrived as fighting men sought every possible scrap of knowledge about the winds, oceans, and beaches where they would do battle. This paid off as meteorologists and other geoscientists provided life-or-death information for everything from bombing missions to the Normandy invasion.

In 1945, as the war effort wound down, scientists wondered what would become of these enterprises. The U.S. Navy decided to step in and fund basic research through a new Office of Naval Research. The support for science, later taken up by other military services as well, was propelled by a group of officers who saw they would need scientists for many purposes. The war had been shortened, if not decided, by radar, atomic bombs, and dozens of other scientific devices barely imagined a decade earlier. Who could guess what basic research might turn up next? Ready access to skilled brains might be vital in some future emergency. Meanwhile, scientists who made famous discoveries would bring prestige to the nation in the global competition with the Soviet Union that was getting under way—the Cold War. So there was reason to support good scientists regardless of what questions they chose to pursue. Still, some fields of science were more equal than others in the long-term advantages they might provide to the United States.

Physical geoscience was one of the privileged fields. Military officers recognized that they needed to understand almost everything about the environments in which they operated, from the ocean depths to the top of the atmosphere. In view of the complex interconnectedness of all things geophysical, the military services were ready to sponsor many kinds of research. For good practical reasons, then, the U.S. government supported geophysical work in the broadest fashion. If purely scientific discoveries happened along the way, they would be a welcome bonus.

Meteorology was especially favored. The U.S. Air Force had a natural concern for the winds and was particularly generous with support, but other military and civilian agencies joined to foster research that might eventually improve weather prediction. Beyond the daily forecast, some experts had visions of deliberately altering the weather. Schemes to provoke rain by "seeding" clouds with silver iodide smoke caught the public's attention in the 1950s, and government officials and politicians took heed. The U.S. government was pressed to fund a variety of meteorological studies in hopes of improving agriculture with timely rains. A nation that understood weather might also obliterate an enemy with droughts or endless snows—a Cold War indeed! A few scientists warned that "climatological warfare" from cloud seeding or the like could become more potent than even nuclear bombs.

These programs all addressed questions of how to predict, and perhaps control, the weather temporarily within a localized region. Questions about long-term climate change over the planet as a whole were *not* a favored field of inquiry. Why pay for research about, for example, the global effects of increased CO_2, when such change was not expected for centuries to come, or more likely never?

Nobody advised Gilbert Plass to study greenhouse warming. The Office of Naval Research supported his work of theoretical calculations for an experimental group at Johns Hopkins University that was studying infrared radiation. As Plass later recalled, he got curious about climate change only because he read broadly about topics in pure science. He happened on the discredited theory that the ice ages could be explained by changes in CO_2. As an adjunct to his official work, Plass took to studying how CO_2 in the atmosphere absorbed infrared radiation. Before he finished his analysis, he moved to southern California to join a group at the Lockheed Aircraft Corporation that was studying questions of infrared absorption directly related to heat-seeking missiles and other weaponry. Meanwhile, he wrote up his results on the greenhouse effect—"in the evening," as he recalled, taking a break from his weapons research.[2]

Plass knew the old objection to the greenhouse theory of climate change: in the parts of the spectrum where infrared absorption took place, the CO_2 and water vapor that were already in the atmosphere sufficed to block all the radiation that could ever be blocked, so a change in the level of the gas could not matter. Doubts about this had been raised during the 1940s by new measurements and an improved theoretical approach. In fact, the old measurements had been faulty; the CO_2 absorption bands are not totally saturated, and they do not entirely overlap the water vapor bands. More important, a change in gas levels would matter even if all heat radiation was already blocked in the lower atmosphere. After all, the surface of the planet would heat up without limit unless it somehow shed the energy received from the Sun. The heat energy must work its way up through the atmosphere, layer by layer, until it reaches the thin upper layers where radiation does escape easily into space. In the frigid and rarefied air up there, the broad bands that tended to block radiation at sea level resolved into clusters of narrow spectral lines, like a picket fence with spaces where radiation could slip through. Plass realized that adding more CO_2 in an upper layer would indeed make a difference. As in Tyndall's analogy of a dam that makes the river behind it rise until it is high enough to get over, the extra gas would increase the blockage that forced a rise in temperature in all the layers below.

Plass could say nothing more specific without extensive computations. Fortunately, he had access to the newly invented digital computers. His lengthy calculations demonstrated that adding or subtracting some CO_2 could make a difference, seriously reducing or increasing the amount of radiation that escaped from the Earth's surface into space. In 1956 Plass announced that human activity would raise the average global temperature "at the rate of 1.1 degree C per century."[3]

Plass's computation was too crude to convince other scientists, for he had left out crucial factors such as possible changes in water vapor and clouds. But he did prove a central point: the greenhouse effect could not be dismissed with the old argument that adding more CO_2 could make no difference. He warned that climate change could be "a serious problem to future generations"—although not for several centuries. Like Arrhenius and Callendar, Plass was chiefly interested in the mystery of the ice ages. If the world's temperature continued to rise to the end of the century, he thought it would matter mainly to scientists, by confirming the CO_2 theory of climate change.[4]

Lockheed was not far from the California Institute of Technology, where Dave Keeling was pursuing his own questions about CO_2. He read Plass's work, talked with him, and was impressed. When Keeling had begun studying the fluctuating levels of CO_2 in the atmosphere, he had spoken of possible applications for agriculture. But his true interest was the pure, scientific study of geochemistry on a global scale. What processes affected the level of CO_2, and what did that level affect in return?

Answering such questions would need measurements at a level of accuracy beyond what could be reached by any instrument on the market. Keeling spent months of research and labor building his own instrument. As he measured the air in various locations around California, patiently refining his techniques, he found that at the most pristine locations he kept getting the same number. It must be the true base level of atmospheric CO_2, underlying the passing pulses emitted by an upwind factory or farm. (The Scandinavian scientists who monitored the gas never hoped to see any such thing, and they failed to hunt out all the errors in their techniques, which Keeling managed to do.)

The next question was whether the level was gradually rising, as Plass and Callendar suspected. That question had little chance to attract research funds. Experts believed any rise of CO_2 would be too slow to matter for a long time to come, and it probably couldn't happen at all. For if Plass had shown that the facts of infrared absorption did not rule out greenhouse warming, another weighty objection to the theory remained. Wouldn't the oceans simply swallow up whatever extra CO_2 we humans might pour into the atmosphere?

It happened that the movements of carbon could now be tracked with a new tool, radiocarbon—a radioactive isotope, carbon 14. Such isotopes had come under intense study during the wartime work to build nuclear weapons, and the pace had not slackened in the postwar years. Sensitive instruments were developed to detect radioactive fallout from Soviet nuclear tests, and a few scientists turned the devices to measuring radiocarbon. Their studies also drew support from interests far removed from the Cold War. Archeologists and the philanthropists who supported them were fascinated by the way radiocarbon measurements could give exact dates for ancient relics such as mummies or cave bones. The isotope is created in the upper atmosphere, when cosmic-ray particles from outer space strike nitrogen atoms and transform them into radioactive carbon. Some of the radiocarbon finds its way into living creatures. After a creature's death the isotope slowly decays over millennia at a fixed rate. Thus, the less of it that remains in an object, in proportion to normal carbon, the older the object is.

One of the new radiocarbon experts, the chemist Hans Suess, thought of applying the technique to the study of geochemistry. It occurred to him that the carbon emitted when humans burned fossil coal and oil is ancient indeed, its radioactivity long gone. He gathered wood from century-old trees and compared it with modern samples. In 1955 Suess announced that he had detected that ancient carbon had been added to the modern atmosphere, presumably from the burning of fossil fuels. But he figured that the added carbon made up barely 1 percent of all the carbon in the atmosphere—a figure so low that he concluded that most of the carbon derived from fossil fuels was being promptly taken up by the oceans. A decade would pass before he managed to get more accurate measurements, which would show a far higher fraction of fossil carbon.

Everyone knew that radiocarbon measurements were tricky; Suess's data were obviously preliminary and uncertain. The important thing he had demonstrated was that fossil carbon had shown up in the atmosphere. With more work one might figure out exactly how long it would take the oceans to absorb the carbon derived from burning fossil fuels. The question was intriguing, for as an oceanographer admitted, "Nobody knows whether it takes a hundred years or ten thousand."[5]

This oceanographer was Roger Revelle, a dynamo of a researcher and administrator who was driving the expansion of the Scripps Institution of Oceanography near San Diego, California. Sitting on a dramatic cliff overlooking the Pacific, the prewar Scripps had been a typical oceanographic establishment, quiet and isolated, with its small clique of a dozen or so gossiping researchers and a single research ship. It relied on private patronage, which faltered when the Depression bit into the Scripps family's funds. The postwar Scripps was growing into something quite different, a complex of modern laboratories. Revelle, supplementing the basic support that he got from the public purse through the University of California, won funding for a variety of projects under contracts from the Office of Naval Research, the natural patron for any research related to the oceans, and other federal agencies. One of Revelle's many good ideas was to use some of the money to hire Suess to come to Scripps and pursue radiocarbon studies. By December 1955 the two had joined forces, combining their expertise to study carbon in the oceans.

From measurements of radiocarbon in seawater and air, Suess

and Revelle deduced that the ocean surface waters took up a typical molecule of CO_2 from the atmosphere within a decade or so. Other scientists who looked into the question around the same time confirmed the conclusion. Yes, the oceans were absorbing most of the carbon humanity added to the atmosphere. Revelle and Suess wrote this up for publication.

The only question left, it seemed, was whether it would accumulate near the surface, or whether currents would carry it deep into the oceans. Revelle's group was already studying the question of how fast the ocean surface waters turned over. It was a matter of national interest, for the navy and the U.S. Atomic Energy Commission were concerned about the fate of fallout from bomb tests. The Japanese were in an uproar over contamination of the fish they relied on. Moreover, if the ocean currents were slow enough, radioactive waste from nuclear reactors might be dumped on the seabed. Measurements of radiocarbon at various depths and other studies, pursued at Scripps and elsewhere in the 1950s, showed that on average the ocean waters turn over completely in several hundred years. That seemed fast enough to sweep the CO_2 produced by human industry into the depths.

Revelle liked to pursue various lines of research in parallel, and another of his interests happened to be relevant. Back in 1946 the navy had sent Commander Revelle (he had taken the naval rank during the war) to Bikini Atoll, at the head of a team that studied the lagoon in preparation for a nuclear bomb test. The seas are not just salt water but a complex stew of chemicals, and it was not easy to say what any change would mean for, say, the carbon in coral. Through the following decade Revelle kept trying out calculations and abandoning them, until one day he realized that the peculiar chemistry of seawater would prevent it from retaining all the carbon that it might take up.

The mix of chemicals in seawater creates what chemists call a buffering mechanism, which stabilizes the water's acidity. This had been known for decades, but nobody had realized what it meant for CO_2. Now Revelle saw that when some molecules were absorbed, their presence would alter the chemical balance through a chain of reactions. Although it was true that most of the CO_2 molecules added to the atmosphere would wind up in ocean surface water within a few years, most of these molecules (or others already in the oceans) would promptly be evaporated back out. Revelle calculated that in sum, the ocean surface could not really absorb much gas— barely one-tenth the amount his earlier calculations had predicted.

It was now 1957, and Revelle and Suess were ready to send off for publication their paper describing how the oceans would promptly absorb all the extra CO_2 humanity could produce. Revelle went back and added a few sentences to explain why this would not in fact occur. As sometimes happens with landmark scientific papers, written in haste while understanding just begins to dawn, Revelle's self-contradictory discussion was so obscure that it took other scientists a couple of years to understand and accept his discovery. Revelle himself did not grasp all the implications at first. In the revised paper he included a quick calculation that the amount of CO_2 in the atmosphere would rise gradually over the next few centuries, then level off with a total increase of 40 percent or less.

That reassuring conclusion was a gross underestimate. Revelle was assuming that industrial emissions during the coming centuries would continue at the 1957 rate. Scarcely anyone had yet grasped the prodigious fact that both population and industrialization were exploding in exponential growth. Between the start of the twentieth century and its end, the world's population would quadruple, and the use of energy by an average person would quadruple, making a sixteen-fold increase in the rate of emission of CO_2. Yet at midcentury, world wars and the Great Depression had led most technologically advanced nations to worry about a possible *decline* in their populations. Their industries seemed to be plodding ahead, expanding no faster in that decade than in the last. As for "backward" regions like China and Brazil, industrialization had scarcely entered anyone's calculations except as a possibility for the remote future.

Different ideas were beginning to stir. In particular, Keeling's mentor at Caltech, the geochemist Harrison Brown, was sketching out a more realistic vision of a future with exploding population and industrialization. Revelle had heard these ideas, and before he sent off for publication the paper he had written with Suess, he added a remark: the accumulation of CO_2 "may become significant during future decades if industrial fuel combustion continues to rise exponentially." In conclusion, he wrote, "Human beings are now carrying out a large scale geophysical experiment of a kind that could not have happened in the past nor be reproduced in the future."[6]

Revelle meant "experiment" in the traditional scientific sense, a nice opportunity for the study of a geophysical process. Yet he did recognize that there might be some future risk. Other scientists too began to feel a mild concern as they gradually assimilated the meaning of Plass's and Revelle's difficult calculations. Adding CO_2 to the atmosphere could change the climate after all. And the changes might arrive not in some remote science-fiction future, but within the next century or so.

Revelle and Suess, like Arrhenius, Callendar, Plass, and everyone else who up till then had made a contribution toward the discovery of global warming, had taken up the question as a side issue. They saw in it a chance for a few publications, a detour from their main professional work, to which they soon returned. If just one of these men had been possessed by a little less curiosity, or a little less dedication to laborious thinking and calculation, decades more might have passed before the possibility of global warming was noticed. It was also a historical accident that military agencies were scattering money with a free hand in the 1950s. Without the Cold War there would have been little funding for research on a subject nobody had connected with practical affairs. The U.S. Navy had bought an answer to a question it had never thought to ask.

Revelle and Suess were now eager to learn more about the "large scale geophysical experiment" of greenhouse warming. Few others paid much attention to their difficult technical paper. Government agencies like the navy could hardly be expected to devote much more money to a question that seemed unlikely to yield anything useful, or even more scientific information, without great effort. Fortunately, just then another purse opened up. The new funds came (or seemed to come) from peaceful internationalism, altogether apart from national military drives.

Geophysics is inescapably international. Ocean currents and winds are ignorant of borders. Yet up through the mid-twentieth century, most geophysicists studied phenomena within a region, often not even an entire nation but part of a nation. Meteorologists of different nationalities did cooperate in the loose informal fashion of all science, reading one another's publications and visiting one another's universities. Gradually, their subject drove them to join forces more closely. In the second half of the nineteenth century, leaders of the field got together in a series of international congresses, which led to the creation of the International Meteorological Organization. Similar organizational drives were strengthening other fields important for climate. In 1919 the International Union of Geodesy and Geophysics was founded, with sections for fields such as oceanography. Still, most individuals did the bulk of their work within the confines of a single field such as geology or meteorology and rarely met their foreign colleagues.

After the Second World War, governments saw new reasons to support international cooperation in science. This was the era that saw the creation of the United Nations, the Bretton Woods financial institutions, and many other multilateral efforts. The aim was to bind peoples together with interests that transcended the self-serving nationalism that had brought so much horror and death.

When the Cold War began, it only strengthened the movement, for if tens of millions had recently been slaughtered, nuclear arms could slay hundreds of millions. It seemed essential to create areas where cooperation could flourish. Science, with its long tradition of internationalism, offered some of the best opportunities. Fostering transnational scientific links became an explicit policy of the world's leading democracies, not least the United States. It was not just that gathering knowledge gave a handy justification for creating international organizations. Beyond that, the ideals and methods of scientists, their open communication, their reliance on objective facts and consensus rather than command—all would reinforce the ideals and methods of democracy. As the political scientist Clark Miller explained, American foreign policy makers believed the scientific enterprise was "intertwined with the pursuit of a free, stable, and prosperous world order."[7]

Study of the global atmosphere seemed a natural place to start. In 1947 a convention of experts and government representatives explicitly made weather studies an intergovernmental enterprise. In 1951 the International Meteorological Organization was replaced by the World Meteorological Organization, or WMO (one of the first of the acronyms that would infest everything geophysical). The WMO, an association of national weather services, soon became an agency of the United Nations. That gave meteorological groups access to important organizational and financial support and lent them a new authority and stature.

All this did little to connect the scattering of scientists in diverse fields who might say something about climate change. Most turned their attention for only a few years to some special aspect of the topic. An astrophysicist studying changes in solar energy, a geochemist studying the movements of radioactive carbon, and a meteorologist studying the global circulation of winds had little in common. They were unlikely to meet at a scientific conference, read the same journals, or even know of one another's existence.

Revelle's decision to bring Suess to Scripps and join forces, ocean-ographer with geochemist, was a well-conceived exception to the rule.

By the mid-twentieth century, few scientists managed to carry out significant work in more than one field. The knowledge needed had become too deep, and the techniques too esoteric. Trying to be-come expert in a second field of knowledge sucked up energy and risked one's career. "Entering a new field with a degree in another is not unlike Lewis and Clark walking into the camp of the [Native American] Mandans," remarked Jack Eddy, a solar physicist who took up the study of climate change. "You are not one of them . . . Your degree means nothing and your name is not recognized. You have to learn it all from scratch."[8]

To make communication still harder, different fields attracted different kinds of people. If you went into the office of a statistical climatologist, you could expect to find ranges of well-organized shelves and drawers stacked with papers bearing neat columns of figures. In later years the stacks would hold computer printouts, the fruit of countless hours spent coding programs. The climatologist was probably the kind of person who, as a boy, had set up his own home weather station and meticulously recorded daily wind speed and rainfall, year after year. Go into the office of an oceanographer and you were more likely to find a jumble of curiosities from the shores of the seven seas. You could hear adventure stories, like one experienced scientist's tale of how he was washed overboard and es-caped drowning by a hair. Oceanographers tended to be salty types, accustomed to long voyages far from the comforts of home, self-reliant and outspoken.

These differences went along with divergence in matters as fun-damental as the sorts of data people used. Climatologists, for exam-ple, relied on the WMO's world-spanning network of thousands of weather stations where technicians reported standardized data. Oceanographers personally built their instruments and lowered them over the side of one of their few research ships. The climatologist's weather, constructed from a million numbers, was something entirely different from the oceanographer's weather—a horizontal blast of sleet or a warm, relentless trade wind. As one climate expert remarked in 1961, "The fact that there are so many disciplines involved, as for instance meteorology, oceanography, geography, hydrology, geology and glaciology, plant ecology and vegetation history—to mention only some—has made it impossible to work . . . with common and well established definitions and methods."[9]

Such fragmentation was becoming intolerable. In the mid-1950s a small band of scientists worked out a scheme to boost cooperation among the various geophysics disciplines. They hoped to coordinate data gathering on an international scale and—no less important—to persuade governments to add an extra billion or so dollars to their funding of geophysics research. Their plans culminated in the creation of the International Geophysical Year of 1957–1958. The IGY would draw together scientists from many nations and a dozen different disciplines, to interact in committees that would plan and carry out interdisciplinary research projects grander than any attempted before.

A variety of motives converged to make the IGY possible. The government officials who supplied the money, while not indifferent to pure scientific discovery, expected the new knowledge would have civilian and military applications. The American and Soviet governments and their allies further hoped to win practical advantages in their Cold War competition. Under the banner of the IGY they could not only enhance their national prestige, but also collect data of potential military value. Some scientists and officials, conversely, hoped that the IGY would help set a pattern of cooperation between the rival powers—as indeed it did. It is a moot question whether, in a more tranquil world, governments would have spent so much to learn about seawater and air. For whatever motives, the result was a coordinated effort involving several thousand scientists from sixty-seven nations.

Climate change ranked low on the list of IGY priorities. But with such a big sum of new money, there was bound to be something for climate-related topics. The study of CO_2 was one minor example. In the committees that allocated the U.S. share of funding, Revelle and Suess argued for a modest program to measure the gas in the ocean and air simultaneously at various points around the globe. It wouldn't cost much, so the committee granted some money. Revelle already had Keeling in mind for this work, and he now hired the young geochemist to come to Scripps and conduct the world survey. Revelle aimed to establish a baseline "snapshot" of CO_2 levels around the world, averaging over the large variations observed from place to place and from time to time. After a couple of decades, somebody could come back, take another snapshot, and see if the average CO_2 level had risen.

Keeling aimed to do better than that. "Keeling's a peculiar guy," Revelle later remarked. "He wants to measure CO_2 in his belly . . . And he wants to measure it with the greatest precision and the greatest accuracy he possibly can."[10] The greatest accuracy called for expensive new instruments, far more precise than most experts thought were called for to measure something that fluctuated as widely as CO_2 levels. Keeling lobbied key officials and managed to persuade them to give him money for the instruments. He set one up atop the volcanic peak Mauna Loa in Hawaii, surrounded by thousands of miles of clean ocean, one of the best sites on Earth to measure the undisturbed atmosphere. Another instrument went to the even more pristine Antarctic.

Keeling's costly equipment, together with his relentless pursuit of every possible source of error, paid off. In Antarctica, he tracked down variations in the CO_2 measurements to emissions from nearby machinery. On Mauna Loa, gas leaking from vents in the volcano itself was to blame. Stalking such problems with meticulous attention to detail, Keeling nailed down a remarkably precise and consistent baseline number for the level of CO_2 in the atmosphere. His first twelve months of data hinted that a rise could be seen in just that one year.

But the IGY was winding down. By November 1958 the remaining funds had fallen so low that CO_2 monitoring would have to stop. Keeling scrambled to find more money. Suess and Revelle diverted a fraction from a grant that the Atomic Energy Commission had given Scripps for other purposes. (In those days, more than now, agencies trusted scientists to spend funds as they chose.) In 1960, with two full years of Antarctic data in hand, Keeling reported that the baseline CO_2 level had risen.[11] The rate of the rise was approximately what would be expected if the oceans were not swallowing up most industrial emissions (Figure 1).

Although some scientists immediately recognized the importance of Keeling's work, no agency felt responsible for funding a climate study that might run for many years. In 1963 the work almost had to shut down. Keeling turned to the National Science Foundation, a U.S. federal agency established back in 1950 on a modest budget. The 1957 launching of the Soviet *Sputnik* satellite had spurred Americans, fearful of trailing behind their Communist enemy, to boost funding for all areas of science. The NSF, its wallet fattened, took over from military agencies much of the nation's support of basic research. One minor consequence was that Keeling got funds to continue the Mauna Loa measurements.

As the Mauna Loa data accumulated, the record grew increasingly impressive, showing CO_2 levels noticeably higher year after year. What had begun as a temporary job for Keeling was turning into a lifetime career—the first of many careers that scientists would eventually dedicate to climate change. Within a few years, Keeling's inexorably rising CO_2 curve was widely cited by scientific review panels and science journalists. It became the central icon of the greenhouse effect.

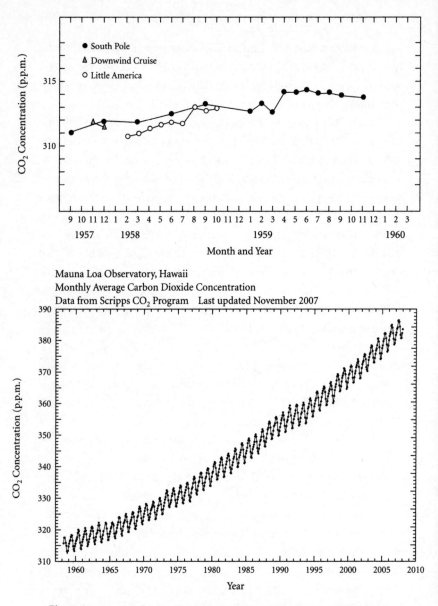

Mauna Loa Observatory, Hawaii
Monthly Average Carbon Dioxide Concentration
Data from Scripps CO_2 Program   Last updated November 2007

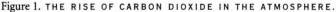

Figure 1. THE RISE OF CARBON DIOXIDE IN THE ATMOSPHERE.

Upper: A rising level of CO_2 in the atmosphere was first demonstrated in 1960 in Antarctica, visible after only two years of measurements. (C. D. Keeling, *Tellus* 12 [1960]: 200, reproduced by permission.) *Lower:* The "Keeling curve" of CO_2 measured at Mauna Loa, Hawaii, through five decades. Within the long-term rise are annual fluctuations as Northern Hemisphere plants take up carbon during summer growth and release it in winter decay. (Scripps Institution of Oceanography, reproduced by permission.)

Keeling's data put a capstone on the structure built by Tyndall, Arrhenius, Callendar, Plass, and Revelle and Suess. This was not quite the discovery of global warming. It was the discovery of the *possibility* of global warming. Experts would continue for many years to argue over what would actually happen to the planet's climate. But no longer could a well-informed scientist dismiss out of hand the possibility that our emission of greenhouse gases would warm the Earth. That odd and unlikely theory now emerged from its cocoon, taking flight as a serious research topic.

CHAPTER THREE

A DELICATE SYSTEM

The weather means a lot to people in Boulder, Colorado, a city of mountaineers and skiers, including not a few scientists. The winter winds there can nudge cars off the road, and in summer you can sit in the hills above town to watch thunderstorms sail across the high plains, churning with lightning. One of the most striking sights in Boulder is the sandstone-red towers of the National Center for Atmospheric Research. Created by Congress in 1960 under pressure from scientists seeking a way around the Weather Bureau's cobwebs, NCAR was dedicated to the scientific study of everything in the atmosphere from cloudbursts to climate. Boulder was a good spot for the conference titled "Causes of Climate Change" that convened there in August 1965. The meeting was scarcely noticed by most scientists at the time, but in retrospect it was a turning point.

The organizers had deliberately brought together experts in everything from volcanoes to sunspots, with an oceanographer, Roger Revelle, to preside over them. Spirited debate filled the lectures and roundtable discussions as rival theories clashed, and Revelle needed all his exceptional leadership skills to keep the meeting on track. The conference had convened mainly to discuss the many rival explanations of the ice ages in the comfortable, traditional mode. Instead, it exploded with new ideas that pointed to a novel and foreboding way of looking at the future of climate. The planet's climate, the scientists agreed, could not be treated in the old fashion, like some simple mechanism that kept itself stable. It was a complex system, precariously balanced. The system showed a dangerous potential for dramatic change, on its own or under human technological intervention, and quicker than anyone had supposed.

At the Boulder meeting not only the climate, but ways of studying it, appeared in a new light. The familiar unchanging climatology of statistical compilations held no appeal for these scientists. They were trying to build up their knowledge from solid mathematics and physics, aided by new techniques drawing on fields from microbiology to nuclear chemistry. But science alone could not explain the deep shift in views about one of the fundamental components of human experience. Events had altered the thinking of everyone in modern society.

Is human technology a force of geophysical scope, capable of affecting the entire globe? Surely it is not, thought most people in 1940. Surely it is, thought most in 1965. The reversal resulted from visible connections between technology and the atmosphere. One of these was a growing awareness of the dangers of atmospheric pollution. In the 1930s citizens had been happy to see smoke rising from factories, for dirty skies meant jobs. But in the 1950s, as the economy soared and life expectancy lengthened in industrialized countries, a historic shift began: fewer worries about poverty, more worries about chronic health conditions. Doctors were learning that air pollution was dangerous. A "killer smog" that smothered London in 1953 demonstrated that the stuff we put into the air could actually slay several thousand people in a few days. Meanwhile, adding to the smoke from coal burning came exhaust from rapidly proliferating automobiles.

Also drawing public attention to interference with the air, people claimed they could make rain by "seeding" clouds. The promises of weather control did not make everyone happy. When one or another community attempted to bring rain to its fields, people downwind might hire lawyers to complain that they had been robbed of their precipitation. Meanwhile, scientists openly speculated about other technical tricks, such as spreading a cloud of particles at a selected level in the atmosphere to interfere with solar radiation. Journalists and science-fiction writers suggested that with such techniques, the Russians might someday inflict terrible droughts or blizzards on the United States. It had become plausible that by putting materials into the air, humans could alter climate on the largest scale—and not necessarily for the better.

The biggest stimulus to changes in thinking was the astonishing advent of nuclear energy. Suddenly nothing seemed beyond human power. Experts speculated that salvoes of atomic bombs could, among other wonders, control weather patterns, bringing rain just where it was needed, or even warming the Arctic wastes to make them habitable. More common than utopian dreams, however, were apocalyptic nightmares. Movies and novels pictured the extinction of all life by radioactive fallout, carried around the world on the winds after a nuclear war.

Exquisitely sensitive instruments could detect radioactive fallout from nuclear test explosions half a world away—the first recognized form of global atmospheric pollution. And in 1962 Rachel Carson published *Silent Spring,* warning that pesticides such as DDT and other chemical pollution, drifting around the world much like fallout, could endanger living creatures not just in the neighborhood of the polluter, but everywhere. Feelings of dread multiplied: whether or not technology would turn deserts into gardens, it could demonstrably turn gardens into deserts!

Many among the public suspected that radioactive dust from bomb tests was already affecting the weather. From about 1953 until open-air testing ceased in the mid-1960s, some people blamed the faraway explosions for almost any unseasonable heat or cold, drought or flood. In a magazine article laying out the evidence that global temperatures were rising, the authors remarked, "Large numbers of people wonder whether the atomic bomb is responsible for it all."[1]

The new threats awoke images and feelings normally confined to nightmares. Humans were meddling with the very winds and rain, polluting everything. Would Mother Nature exact revenge for our unhallowed attacks on her? Ancient myths of how human sins against the natural order brought afflictions, all the way up to a fiery end of the world, took on a veneer of scientific plausibility. The charged language of defilement and transgression rarely appeared outside the nuclear and chemical controversies. But for climate, too, many felt that natural processes somehow followed not only scientific law but also moral law. Since prehistoric times, tribal peoples had attributed climate troubles such as an unusually bad winter to the violation of a taboo, and in modern times it was not unknown for preachers to beg forgiveness in their prayers for rain.

Once people began to grasp that human technology actually could affect the entire planetary system, for better or worse, journalists found it easier to suggest that burning fossil fuels could change the climate. Evidence that the world had been growing measurably warmer had become strong enough to convince most meteorologists. During the 1950s newspaper readers repeatedly saw small items reporting anecdotes of warming, especially in the Arctic. For example, in 1959 the *New York Times* reported that the ice in the Arctic Ocean was only half as thick as it had been in the previous century. Still, the report concluded, "the warming trend is not considered either alarming or steep."[2] Compared with chemical pollution and nuclear war, climate change was an old-fashioned issue, and most likely harmless.

Revelle took the lead in suggesting that trouble might lie ahead. As soon as he calculated that a rise in the CO_2 level was likely, Revelle took pains to talk about greenhouse warming with science journalists and government officials. Noting that climate had changed abruptly in the past, perhaps bringing the downfall of entire civilizations in the ancient world, he warned that the CO_2 greenhouse effect might turn southern California and Texas into "real deserts." Testifying before Congress in 1956 and 1957, he was one of the first to use a new and potent metaphor: "The Earth itself is a space ship," he said. We had better keep an eye on its air control system.[3]

Revelle did not expect much change in the climate for many decades, and perhaps never. He meant only to prod the government to cough up funds for the IGY and other geophysical research. Few others foresaw anything ominous at all. "There would seem to be every reason for producing as much carbon dioxide as we can manage," one popularization concluded. "It is helping us towards a warmer and drier world."[4] In any case, nothing would happen until the twenty-first century—and that was so far away! The subject was scarcely noticed by anyone outside the science-minded minority who happened on reports by science journalists. Those were mostly buried in the back pages of newspapers or dropped into news magazines as a brief paragraph or two.

The one stubborn fact that could not be overlooked was the climbing level of CO_2 in the atmosphere. Keeling's curve rose year by year through the 1960s. A few scientists began to think that the matter might be so important that people should consider taking action. By this they meant pursuing climate research more energetically, to find the actual dimensions of the issue.

The first step in that direction came in 1963, when Keeling and a few other experts met at a conference sponsored by the private Conservation Foundation. They issued a report suggesting that the doubling of CO_2 projected for the next century could raise the world's temperature by 4°C (more than 6°F). They warned that this could be harmful; for example, it could cause glaciers to melt and raise the sea level enough to submerge important coastal areas. The federal government should give the subject more consistent attention, they said, with better organization and more money.[5]

The government reacted to such complaints at its usual deliberate pace. In 1965, when the President's Science Advisory Committee formed a panel to address environmental issues, it included a subpanel of climate experts. They reported that greenhouse-effect warming was a matter of real concern. That put the issue on the official agenda at the highest level—although only as one item on a long list of environmental problems, many of which seemed more pressing. The next step in such matters was typically to ask the National Academy of Sciences to form a committee and issue an authoritative report. In 1966 the Academy duly pronounced on how human activity could influence climate. The experts sedately said there was no cause for dire warnings, but they did believe the CO_2 buildup should be watched closely. "We are just now beginning to realize that the atmosphere is not a dump of unlimited capacity," the report said, "but we do not yet know what the atmosphere's capacity is."[6] The primary conclusion was typical of such reports—a maxim that came from the heart of scientists' belief in their calling—More Money Should Be Spent on Research.

That was easier said than done. As the 1966 Academy report explained, for climate as for most environmental fields, "there exists no single natural advocate in the Federal structure, nor is there a clear mechanism for making budgetary decisions." In the mid-1960s a scattering of government agencies together spent roughly $50 million a year for all aspects of meteorological research. That was not much, and climate change caught only a small fraction of it.[7] Other nations were spending even less.

Agencies gave priority to predicting weather a few days ahead. When thoughts turned to longer terms, the titles of the panels show what was on people's minds. The president's advisory group was named the Environmental Pollution Panel, and the Academy's was the Panel on Weather and Climate Modification. Asked about human influence on the atmosphere, the public would think first about smog, and next about deliberate attempts to make rain. Progress on global warming would come mainly from money earmarked for other purposes.

In climate studies the great intellectual prize, still unclaimed after a century of work, was the explanation of the ice ages. Hope for progress lay in improved techniques. The most precise and ingenious one had been invented early in the century by a few Swedish scientists: the study of ancient pollens. The tiny but amazingly durable pollen grains are as various as sea shells, with baroque lumps and apertures characteristic of the type of plant that produced them. One could dig up soil from lake beds or peat deposits, dissolve away in acids everything but the sturdy pollen, and after some hours at a microscope know what kinds of flowers, grasses, or trees had lived in the neighborhood at the time the layer was formed. That told scientists much about the ancient climate. We had no readings from rain gauges and thermometers 50,000 years ago, but pollen served as an accurate "proxy."

The usefulness of pollen and other proxies for ancient climate redoubled when radiocarbon dating arrived. Go to a glacial moraine or lake bed, dig out fragments of ancient trees, measure the fraction of radioactive carbon in them, and you had a date. Gradually scientists worked out a reliable timescale for climate fluctuations. For example, dating of lake deposits in the western United States showed highly regular cycles of drought and flood. It was hard to match these with the geologists' traditional sequence of four ice-age cold and warm periods. Rather, they seemed to match the 21,000-year cycle in Milankovitch's astronomical calculations.

Another promising new technique likewise came from a clever combination of biology and nuclear science. The oceans swarm with microscopic plankton, including countless foraminifera (nicknamed forams), single-celled animals that scavenge with pseudopods poking out through holes in their shells. When forams die, their tiny shells drift down into the ooze of the seabed and there endure for ages, in some places so numerous that they form thick deposits of chalk or limestone. In 1947 the nuclear chemist Harold Urey found a way to take the temperature of an ancient ocean by measuring the oxygen that forams built into their shells. The rare isotope oxygen 18 is a bit heavier than normal oxygen 16, and biologists had demonstrated that the amount of each isotope that a foram takes up varies with the temperature of the water. The isotopes were fossilized with the shells, and the ratio of isotopes (O^{18}/O^{16}) could be measured with the new and highly sensitive tools developed for nuclear studies.

The idea challenged Cesare Emiliani, a geology student from Italy working in Urey's laboratory at the University of Chicago. There were many problems to solve before he could get reliable results. (Urey, already a Nobel Prize–winner, called it "the toughest chemical problem I ever faced.")[8] And before the chemistry could even begin, samples had to be extracted from the sticky sediments of the ocean floor without disturbing the layers. Börge Kullenberg—let us remember here the name of at least one of the many people who contributed a modest but essential device to this history—solved the problem for a Swedish deep-sea expedition in 1947. He put a piston inside a long tube and pulled the piston up to suck in the sediment while the tube was being shoved into the seabed. Kullenberg could recover cylindrical cores more than 20 meters long. Back in the lab, somebody would put a sample of the muck under a microscope and tease out a few hundred of the shells, each no bigger than the period at the end of this sentence. The shells were ground to powder, which was roasted to extract CO_2 gas so the isotopes could be measured. Technicians needed to be very careful to avoid contamination by any other source of gas, such as their own breath.

In 1955 Emiliani applied these exacting methods to slimy cylinders of mud and clay totaling hundreds of meters in length, extracted from the deep seabed by various expeditions and carefully stored away in oceanographic institutions. Borrowing cores from several expeditions, he patched together a remarkable record of temperature changes stretching back nearly 300,000 years. The rise and fall of temperatures did not seem to match the sequence of ice ages worked out by nineteenth-century geologists. Emiliani finally rejected the entire textbook scheme of four major glacial advances that alternated with long and equable interglacial periods. His data told a tale of dozens of briefer advances and interglacials, complicated by irregular rises and falls of temperature all along the way. His data did correlate rather well with one thing: Milankovitch's complex curve for the amount of summer sunlight at high northern latitudes.

Geologists defended their traditional chronology passionately and skillfully. For a while they held their ground. It turned out that Emiliani's data on oxygen isotopes in foram shells did not directly measure ocean temperatures after all. Instead, as other workers in the late 1960s demonstrated, the isotope ratio changed mainly for another reason. Whenever water evaporated from the oceans and then fell as rain or snow to form continental ice sheets, the ratio of isotopes changed. For a heavier molecule was slightly less likely to evaporate from the ocean surface than a lighter one. Thus, as snowfall built up continental ice sheets, more of the heavier isotope was left behind in the seawater. No matter what the temperature of the water where the forams lived, during a glacial period their shells would wind up with more of the heavier isotope. The changes that Emiliani had detected reflected mainly the changing volume of the planet's ice sheets.

Emiliani defended his results fiercely, loath to admit error. By the end of a decade of debate, all his colleagues had rejected his temperatures. But they were quick to acknowledge that his work, error and all, was still a landmark. If it was the growth and decay of ice sheets that changed the ratio of isotopes, the rises and falls of Emiliani's curves still showed the rhythm of the ice ages.

There was nothing extraordinary about such a combination of error and discovery. Every important scientific paper is written at the outside edge of what can be known. It deserves to be remembered if there is a nugget of value amid the inevitable confusion.

Clear confirmation of Emiliani's basic discovery came when scientists peered through microscopes to take a census of the particular species of foraminifera, counting the differently shaped shells layer by layer down the cores. The assemblage of species varied with the temperature of the water where they had lived. The changes matched Emiliani's curves: the temperatures had indeed changed. Evidently he was right to claim that during the preceding couple of million years there had been dozens of major glaciations—a sequence that roughly agreed with Milankovitch's schedule.

Yet how reliable was any conclusion read from the seabed ooze? As the debate over oxygen isotopes showed, what looks like a simple measured fact can be misleading. Scientists therefore rarely accept a result until it is confirmed by wholly different means. A new voice brought this confirmation, a voice that over the coming decades would increasingly command attention. It belonged to Wallace (Wally) Broecker of the Lamont Geological Observatory. Lamont scientists, isolated amid woods overlooking the Hudson River, were combining geological interests with oceanography and the new radioactive and geochemical techniques in a burst of creative research.

In the 1960s Broecker and a few colleagues had traveled to the tropics to hack samples off ancient coral reefs. Perched at various elevations above the present sea level, the reefs showed how the oceans had risen and fallen as ice sheets built up on the continents and melted away. The reef samples could be dated by measuring their uranium and other radioactive isotopes. These isotopes decayed over millennia on a timescale that had been accurately measured in nuclear laboratories, and, unlike radiocarbon, the decay was so slow that there was still enough left to measure after hundreds of thousands of years. Here, too, Milankovitch's orbital cycles emerged, plainer than ever.

Scientists did not immediately convert to the astronomical-orbit theory. Scientists rarely label a proposed answer to a scientific question "true" or "false," but rather consider how likely it is to be true. Normally, a new body of data will shift opinion only in part, making an idea seem a bit more likely or less likely. At the conference "Causes of Climate Change" held in Boulder in 1965, Broecker would claim only that "the Milankovitch hypothesis can no longer be considered just an interesting curiosity."[9] A few years later, after he and his collaborators accumulated more evidence, they still would not claim to have proved anything for certain. "The often-discredited hypothesis of Milankovitch," they said in 1968, "must be recognized as the number-one contender in the climatic sweepstakes."[10] Some disagreed, reserving the top spot for their own favorite hypotheses.

No matter how strong the match between ancient climate data and orbital cycles, scientists would not find the astronomical theory of climate change really plausible until they had an explanation for how it could physically work. Nobody had forgotten the fundamental objection: the changes in sunlight as the Earth's orbit shifted were trivial. How could a minor change in the forces that bore on the atmosphere make entire continental ice sheets wax and wane? Besides, why should an increase in sunlight at high latitudes in the Northern Hemisphere bring on a worldwide change, melting the ice in the Southern Hemisphere at the same time? If scientists gave serious consideration to Broecker's announcements, that was because other developments were already pushing them to rethink the climate system's fundamental nature.

There had always been meteorologists who challenged common ways of thinking, and some of their speculations had lingered, never quite entirely refuted. For example, back in 1925 a respected climate expert, C. E. P. Brooks, had proposed that ice ages could begin or end almost arbitrarily. He began with the familiar idea that an increase of snow cover would reflect more sunlight, which would further cool the air. Frigid winds, Brooks suggested, would flow into adjoining regions, and the snows would swiftly advance to lower latitudes. Only two stable states of the polar climate were possible, he asserted—one with little ice, the other with a vast white cap on the planet. A shift from one state to the other might be caused by some comparatively slight perturbation. There could be an abrupt and catastrophic rise or fall of tens of degrees, "perhaps in the course of a single season."[11]

Most experts scorned such talk, which seemed only too likely to reinforce notions popularized by religious fundamentalists who were in open conflict with the scientific community. Believers in the literal truth of the Bible insisted that the Earth was only a few thousand years old, and they defended their faith by claiming that ice sheets could form and disintegrate in mere decades.

Scientists replied by pointing to the detailed climate records that were recovered from layers of silt and clay. Analysis showed no changes in less than several thousand years. The scientists failed to notice that most cores drilled from the seabed could not in fact record a rapid change. For the ooze is constantly stirred by burrowing worms, as well as by currents and slumping, which blur any abrupt differences between layers.

Ancient lakes and bogs retained a more detailed record, which scientists could read in the embedded pollen. Major changes in the mix of plants that showed up in a given lake suggested that the last ice age had not ended with a uniformly steady warming, but with some peculiar oscillations of temperature. A particularly striking event, identified in the 1930s in Scandinavia, was a warm period followed by a prolonged spell of bitterly cold weather. The cold spell was dubbed the Younger Dryas after *Dryas octopetala*, an elegant but hardy little Arctic flower whose pollen gave witness to frigid tundra. After that period a more gradual warming followed. In 1955 the timing was pinned down by a radiocarbon study, which revealed that the chief oscillation of temperatures lay around 12,000 years ago. The change had been rapid—where "rapid," for climate scientists at the time, meant something that took as little as one or two thousand years to change. Most scientists thought the event, if real at all, was just a local Scandinavian happenstance.

In 1956 Hans Suess applied his radiocarbon expertise to determine the dates of fossil shells in deep-sea clay drilled up by Lamont Observatory oceanographers. He reported that the last glacial period had ended with a "relatively rapid" rise of temperature—about 1°C per thousand years. Lamont scientists took their own look at the foram shells, sorting them according to whether the species lived in warm or cold waters, and reported an even more abrupt rise. About 11,000 years ago, they said, the climate had shifted from fully glacial conditions to modern warmth within as little as one thousand years. They acknowledged this was "opposed to the usual view of a gradual change."[12]

Emiliani, always prepared to spar with Lamont scientists, disagreed. He published reasons to believe that the temperature rise of some 8°C had been the expected gradual kind, stretching over some 8,000 years. After considerable public debate, he won his point: the variation in the Lamont group's data did not in fact represent a temperature shift. Like some other sudden changes reported in natural records, it reflected peculiarities in the method of analyzing samples, not the real world itself.

Yet in science even mistakes can be valuable if they set scientists to thinking about overlooked possibilities. The apparent evidence that climate had changed drastically in as little as a thousand years had provoked a few people to try to figure out what could have happened. Broecker was at this time still a lowly graduate student at Lamont. He put a bold idea into his doctoral thesis. Looking at the studies that would later be found mistaken, along with a variety of other data, he saw a pattern of glacial oscillations that was very different from the accepted gradual wavelike rise and fall. Broecker suggested that "two stable states exist, the glacial state and the interglacial state, and that the system changes quite rapidly from one to the other."[13] It was just one comment in a thick and little-read doctoral thesis, and it sounded like Brooks's discredited speculations. Most scientists agreed with Emiliani that the fluctuating data, if accurate at all, reflected local oddities. After all, ice sheets kilometers thick would require thousands and thousands of years to build up or melt away. The physics of ice, at least, was simple and undeniable.

A few were not so sure, in particular Broecker's boss—Lamont's founding director, the brash and autocratic Maurice Ewing. Was there any mechanism, he wondered, that could cause climate to swing rapidly between warm and glacial states? Ewing and a colleague, William Donn, had some ideas about how that might actually happen.

If the Arctic Ocean were ever free of ice, argued Ewing and Donn, so much moisture would evaporate that heavy snows would fall all around the Arctic. The snow could build up into continental ice sheets, launching an ice age. But the ice buildup would withdraw water from the world's oceans, and the sea level would drop. Warm currents would no longer flow through the shallow straits into the Arctic Ocean, and it would freeze up. But now the continental ice sheets, deprived of evaporated water, would dwindle away; the seas would rise and eventually send warm currents to melt the Arctic ice pack again. Thus, ice ages would come and go in a grand jumble of feedbacks. Ewing and Donn suggested that the polar ocean might become ice-free and launch us into the next ice age quite soon, as geological time went—maybe even in the next few hundred years.

Scientists promptly contested what many thought was a cockamamie idea. Aside from the standard arguments that ice sheets could not grow so fast, they discovered holes in the Ewing-Donn scheme. For one, they argued, the Arctic straits were sufficiently deep that lowering the sea level would not cut off warm currents. The pair worked to patch up the holes, and for a while many scientists found their arguments at least intriguing. In 1956 a respected U.S. Weather Bureau leader warned that "the human race is poised precariously on a thin climatic knife-edge." If the warming trend that seemed to be under way continued, it might trigger changes with "a crucial influence on the future of the human race."[14] But ultimately the Ewing-Donn scenario won no wider acceptance than most other theories of the ice ages.

Like mistaken data, a mistaken idea can have valuable consequences. Ewing and Donn's model of the ice ages caught the public's attention, offering for the first time an almost respectable scientific backing for images of swift, disastrous climate change. In connection with the reports on how northern regions were growing warmer and ice was retreating, journalists had a fine time exclaiming that the climate might be unstable. For scientists, the daring ideas provoked broad thinking about possible mechanisms for rapid climate change in general. As Broecker later recalled, Donn would "go around and give lectures that made everybody mad. But in making them angry, they really started getting into it."[15]

The new openness to thinking about rapid climate change could get nowhere without better ways of understanding the basic physics of climate. The Ewing-Donn scheme was the last influential climate model in the grand nineteenth-century style, based on what some called "plausibility arguments" and others dismissed as "hand-waving." The only alternative had been mathematical models—trying to catch the pattern of global winds on a page of equations, as a physicist might represent the orbits of planets. But in a century of effort nobody had managed to derive a set of mathematical functions whose behavior approximated that of the real atmosphere.

Starting in the late 1940s, a new way to get a handle on the problem emerged at the University of Chicago, where Carl-Gustav Rossby was encouraging young meteorologists to think like physicists. Dave Fultz built an actual, physical model to test the behavior of the atmosphere. As a simulacrum of the rotating planet, his group put a simple aluminum dishpan on a turntable, using water to stand for the atmosphere. They represented sunlight warming the tropics by heating the dishpan at the outer rim, and they injected dye to reveal flow patterns. The results were exciting, if often mystifying. The crude model showed something rather like the wavering polar fronts that dominate much of the world's weather. Further experiments in Chicago and in Cambridge, England, produced turntable models with something like a miniature jet stream and tiny swirling storms.

Most intriguing of all were photographs that Fultz published in 1959. His rotating dishpan showed a regular circulation pattern that resembled the real world's mid-latitude westerly winds. Stir the water in the dishpan with a pencil, and when it settled down the pattern might have flipped to an altogether different circulation pattern. This was realistic, for the circulation of the actual atmosphere does shift among quite different states; trade winds, for example, come and go with the seasons. Might larger shifts in the circulation pattern cause long-term climate changes?

The dishpan model was only a crude cartoon of the atmosphere. However fascinating it might be, it could not lead to any definite conclusion about our planet. The real contribution of this physical model was its dramatic demonstration that certain systems may be subject to tricky instabilities. Could the real climate shift as abruptly, arbitrarily, and totally as the waves in the dishpan?

The answer would come from an altogether different way of modeling the world, one that had been tried a generation earlier and abandoned as hopeless. In 1922 Lewis Fry Richardson had proposed a complete numerical system for weather prediction. His idea was to divide up a territory into a grid of cells, each with a set of numbers for air pressure, temperature, and so forth, as measured at a given hour. He would then apply the basic physics equations that told how air would respond. He could calculate, for example, a wind speed and direction according to the difference in pressure between two adjacent cells. This would give the pressures and temperatures in the grid cells an hour later, which would serve for the next round of computations, and so forth.

The number of computations was so great that Richardson scarcely hoped his idea could lead to practical weather forecasting. Even if someone assembled a "forecast-factory" employing tens of thousands of clerks with mechanical calculators, he doubted they would be able to compute weather faster than it actually happens. "Perhaps some day in the dim future it will be possible to advance the calculations faster than the weather advances," he wrote wistfully. "But that is a dream."[16] Still, if he could make a numerical model of a typical weather pattern, it would help show how the weather worked.

So Richardson attempted to compute how the weather over Western Europe had developed during a single eight-hour period, starting with the data for a day when there had been coordinated balloon-launchings that measured the atmosphere at various levels. The effort cost him six weeks of pencil work and ended in complete failure. At the center of Richardson's pseudo-Europe, the computed barometric pressure climbed far above anything ever observed in the real world. Taking the warning to heart, for the next quarter century meteorologists gave up any hope of numerical modeling.

What had been hopeless with pencil and paper might possibly be made to work with digital computers. The strange new devices, feverishly developed during the Second World War to break enemy codes and to calculate how to explode an atomic bomb, were leaping ahead in power as the Cold War demanded more and more calculations. In the lead, vigorously devising ways to simulate nuclear weapons explosions, was a brilliant and ambitious Princeton mathematician, John von Neumann. He saw parallels between his nuclear explosion simulations and weather prediction (both involved rapidly changing fluids). In 1946 von Neumann began to advocate using computers for numerical weather prediction.

The idea drew support from the U.S. Weather Bureau, the army, and the air force, as well as the ubiquitous Office of Naval Research. Von Neumann told the navy that his efforts had a dual goal: not only to predict daily weather changes, but also to calculate the general circulation of the entire atmosphere, trade winds and all. That was not because he was interested in global climate change. He and his military sponsors knew they had to understand the general circulation if they were to hope to twist the climate in a given region, to their benefit or an enemy's harm.

To spearhead the project von Neumann recruited Jule Charney, an energetic example of the new generation of mathematical meteorologists from Rossby's group in Chicago. By 1949 Charney's team had computed airflow along a band of latitude. Their results looked fairly realistic—sets of numbers that could almost be mistaken for real weather diagrams, if you didn't look too closely. For example, they could model the effects of a large mountain range on the airflow across a continent. Modeling was taking the first steps toward the computer games that would come a generation later, in which the player acts as a god: raise up a mountain range and see what happens!

The challenge was to find equations that gave plausible results without too many hours of computation. The most famous computers of the 1940s and 1950s were dead slow by comparison with a common laptop computer of later years. With thousands of glowing vacuum tubes connected by tangles of wiring, they ate up a good part of the workers' time repairing the frequent breakdowns. Much of the remaining time was spent devising efficient and realistic mathematical approximations. That called for a rare combination of mathematical ingenuity and physical insight, as well as countless hours of work.

The pages of numbers churned out by a computer run could be validated only by comparison with the characteristics of the actual atmosphere. That called for an unprecedented number of measurements of temperature, moisture, wind speed, and so forth. These were now coming to hand, thanks largely to military needs. During the war and after, nations had set up networks to send up thousands of balloons that radioed back measurements of the upper air. By the 1950s international programs were mapping the daily changes well enough for comparison with the numbers from the rudimentary computer models.

In 1950 Charney's team solved Richardson's problem: for a chosen day they computed something roughly resembling what the weather had actually done. It took them so long to print and sort punched cards that "the calculation time for a 24-hour forecast was about 24 hours, that is, we were just able to keep pace with the weather."[17] By 1955 they had speeded up the process enough to begin issuing actual weather predictions, although it would be another decade before these could consistently beat the traditional map-reading forecasters.

These computer models were regional in scale, not global. But the methods and confidence developed for weather prediction inspired dreams of reaching for the ultimate prize: a model for the general circulation of the planet's atmosphere as a whole. Norman Phillips took up the challenge. With his primitive computer he could calculate no more than one "hemisphere" in the shape of a cylinder, with two levels of atmosphere. Yet by mid-1955 he had devised a numerical model that showed a plausible jet stream and the evolution of a realistic-looking weather disturbance over a few weeks. This settled a long controversy over what processes built the pattern of circulation. For the first time scientists could see, for example, how giant eddies spinning through the atmosphere played a key role in moving energy and momentum from place to place. Phillips's model was quickly hailed as a "classic experiment"—the first true General Circulation Model (GCM).

But Phillips's model eventually exploded. As the calculations stepped forward through time, after twenty or so simulated days the pattern of flow began to look strange. By thirty days the numbers had veered off into conditions never seen on Earth. It looked like the sort of flaw that had destroyed Richardson's project back in 1922. Richardson had thought his calculation would have worked out if only he could have begun with more accurate wind data. But as Rossby pointed out in 1956, people routinely made decent twenty-four-hour predictions by looking at weather maps drawn from very primitive data. "The reasons for the failure of Richardson's prognosis," the puzzled Rossby concluded, "must therefore be more fundamental."[18]

As digital computers proliferated and scientists tried them out on a variety of tasks, results often went oddly astray. The errors might indeed be caused by bad data, as Richardson had thought— "garbage in, garbage out," as computer experts were coming to understand. But the problems might also be introduced by the very nature of numerical computation. For example, computer models had to chop up reality into a grid, where a single average number stood for all the different temperatures in a unit that covered thousands of square kilometers. Modelers spent years devising ways to work around such artificial features in equations and computations.

A few scientists began to wonder whether the exquisite sensitivity of computer models might be saying something about the real world. Start two computations with exactly the same initial conditions and they must always come to precisely the same conclusion. But make the slightest change in the fifth decimal place of some initial number, and as the machine cycled through thousands of arithmetic operations, the difference might grow and grow, in the end giving a seriously different result. Of course it had long been understood that a pencil balanced on its point, for example, could fall left or right depending on the tiniest difference in initial conditions. Most scientists supposed that this kind of situation arose only under exceptionally simplified circumstances, far from the stable balance of huge and complex global systems like climate. Only a few, like Brooks and Ewing, imagined that the entire climate system might be so delicately balanced that a relatively minor perturbation could trigger a big shift.

If that ever did happen, it would be because the perturbation was amplified by feedbacks—a term and a concept that came into fashion in the 1950s. A new science of "cybernetics" was announced by the mathematician Norbert Wiener, whose wartime work on automatic gun-pointing mechanisms had shown him how easily physical systems could wobble out of control. Wiener was at the Massachusetts Institute of Technology, where several groups were enthusiastically building numerical models of weather, among many other things. Wiener advised meteorologists that their attempts were doomed to fail. Quoting the old nursery rhyme that told how a kingdom was lost "for want of a nail" (which caused the loss of a horseshoe of a horse bearing a knight going to a battle), he warned that "the self-amplification of small details" would foil any attempt to predict weather, to say nothing of climate.[19]

In 1961 an accident cast new light on the question. Luck in science comes to those in the right place with the right set of mind, and that was where Edward Lorenz stood. He was at the Massachusetts Institute of Technology, one of the new breed who were combining meteorology with mathematics. Lorenz had devised a simple computer model that produced impressive simulacra of weather patterns. One day he decided to repeat a computation in order to run it longer from a particular point. His computer worked things out to six decimal places, but to get a compact printout he had truncated the numbers, printing out only the first three digits. It was these digits that Lorenz entered back into his computer. After a simulated month or so the weather pattern diverged from the original result. A difference in the fourth decimal place was amplified in the thousands of arithmetic operations, spreading through the computation to bring a totally new outcome.

Lorenz was astonished. He had expected his system to behave like real weather. The truncation errors in the fourth decimal place were tiny compared with any of a hundred minor factors that might nudge the temperature or wind speed from minute to minute. Lorenz had assumed that such variations could lead to only slightly different solutions for the weather a few weeks ahead. Instead, storms appeared or disappeared from the forecast as if by chance.

Lorenz did not shove the puzzle into the back of his mind, but launched into a deep and original analysis. In 1963 he published an investigation of the type of equations that might be used to predict daily weather. "All the solutions are found to be unstable," he concluded. Therefore, "precise very-long-range forecasting would seem to be non-existent."[20] Beyond a few weeks at most, minuscule differences in the initial conditions would dominate the calculation. One calculation might produce rain a week ahead, and the next calculation, fair weather.

That did not necessarily apply to climate, which was an average over many states of weather. Wouldn't the differences in one storm or another balance out, on average, and leave a stable overall result? Lorenz constructed a simple mathematical model of climate and ran it repeatedly through a computer with minor changes in the initial conditions. The results varied wildly. He could not prove that there existed a "climate" at all, in the traditional sense of a stable, long-term statistical average.

These ideas spread among climate scientists, especially at the 1965 Boulder conference "Causes of Climate Change." Lorenz, invited to give the opening address, explained that the slightest change of initial conditions might bring at random a huge change in the future climate. "Climate may or may not be deterministic," he concluded. "We shall probably never know for sure."[21] Meteorologists at the conference also pored over the new evidence that past ice ages might have been started or ended by the almost trivial astronomical shifts that Milankovitch had calculated for the Earth's orbit. Was the climate such a fundamentally unstable system that it could be tipped from one state to another by the slightest push?

That fed into discussions at the conference about the growing suspicion that the climate could change not only more easily but more swiftly than almost anyone had believed a few years earlier. Broecker, Ewing, and others presented a variety of old and new evidence that around 11,000 years ago a truly radical global climate shift had taken place: as much as 5° to 10°C in less than a thousand years. It reminded them of the sudden shifts in the rotating dishpan experiment. To be sure, melting the great ice sheets would necessarily take many thousands of years. But scientists now realized that while that was going on, the less massive atmosphere could fluctuate surprisingly easily. Summing up a consensus at the end of the conference, Revelle declared that minor and transitory changes in the past "may have sufficed to 'flip' the atmospheric circulation from one state to another."[22]

Might the ocean circulation, too, flip between different modes? Some were willing to question the traditional picture of a torpid circulation, unvarying over many thousands of years. Broecker, in particular, had already noticed climate jumps when he was a graduate student and had compared changes in ancient lake levels with ocean data. He had seen signs of drastic changes in ocean cores within less than a thousand years. At the Boulder meeting, Peter Weyl of Oregon State University presented an especially provocative idea. He was developing a complicated theory of the ice ages that involved how changes of saltiness might affect the formation of sea ice. It would scarcely have been noticed among the many other speculative and idiosyncratic models, except for a novel insight. Weyl pointed out that if the North Atlantic around Iceland should become less salty—for example, if melting ice sheets diluted the upper ocean layer with fresh water—the surface layer would no longer be dense enough to sink. The entire circulation that drove cold water south along the sea bottom could lurch to a halt. Without the compensating drift of tropical waters northward, a new glacial period might begin. Others since Chamberlin in the nineteenth century had speculated that a warming of the planet might alter the ocean circulation. Now explicit calculations confirmed that the circulation (what was coming to be called the "thermohaline circulation," from the Greek for heat and salt) was indeed precariously balanced.

Orbital changes, wind patterns, melting ice sheets, ocean circulation—everything seemed to be interacting with everything else. During the 1960s not only climate researchers, but also scientists in other fields and alert members of the public, were coming to recognize that the planet's environment is a hugely complicated structure. Almost any feature of the air, water, soil, or biology might be sensitive to changes in any other feature. Scientists were giving up the traditional approach in which each expert championed a favorite hypothesis about one particular cause of climate change, from volcanoes to the Sun. It seemed more likely that many factors acted together. On top of the external influences, Lorenz and others were pointing out that complex feedbacks could make the system fluctuate under its own internal dynamics. "It is now generally accepted," wrote one authority in 1969, "that most climatic changes . . . are to be attributed to a complex of causes."[23]

This change of viewpoint, toward what some called a holistic approach, pushed scientists to change the way they worked. A plausible model for climate change could not be constructed, let alone checked against data, without information about a great many different kinds of things. It was becoming painfully clear that scientists in a great variety fields needed one another. The 1965 Boulder conference was only one of many occasions when specialists began to interact more closely, drawing on one another's findings or, equally valuable, challenging them.

CHAPTER FOUR

A VISIBLE THREAT

Walter Orr Roberts, an astrophysicist at the University of Colorado, noticed that something was changing in the broad and sparkling skies above Boulder. One morning in 1963, as he was talking with a journalist, he pointed out the jet airplane contrails overhead. He predicted that by mid-afternoon they would spread and thin until you couldn't tell the contrails from cirrus clouds. They did, and you couldn't. Roberts suspected that the airplanes were introducing enough additional cloudiness to affect climate in heavily traveled regions. The human impact on the atmosphere was becoming visible to the discerning naked eye.

Around the same time, a University of Wisconsin meteorologist, Reid Bryson, was flying across India en route to a conference. He was struck by the fact that he could not see the ground, his view blocked not by clouds but by dust. Later he saw similar hazes in Brazil and Africa. The murk was so pervasive that local meteorologists had taken it for granted and never thought to study it. But Bryson realized that the haze was not some timeless natural feature of the tropics: he was seeing smoke from fields set on fire by the growing population of slash-and-burn farmers, and dust from overgrazed pastures turning to desert. Bryson suspected the effects could reach a planetary scale. The smoke and dust, together with growing pollution from industry, might block enough sunlight to cool the Earth significantly. Bryson wrote that he "would be pleased to be proved wrong. It is too important a problem to entrust to a half-dozen part-time investigators."[1]

In truth, scarcely anyone was studying how aerosols—microscopic airborne particles—might influence global climate. Simple physics theory suggested that aerosols should scatter radiation from the Sun back into space, cooling the Earth. Meteorologists had long believed that dust from volcanic eruptions did just that. Aerosols not only intercepted sunlight, but might also affect climate by influencing clouds. Like jet airplane exhaust, all sorts of human emissions might raise cloudiness, adding to the interference with sunlight. Research early in the century had shown that clouds can form only where there are enough "cloud condensation nuclei," tiny particles that give a surface for the water droplets to condense around. Attention to the process redoubled in the 1950s, when "seeding" clouds with silver iodide smoke in hope of making rain became a widespread commercial enterprise.

Studies of cloud seeding with special chemicals in a small locality could not say much about the global impact of the particles emitted by human industry and agriculture. But scarcely anyone attempted to attack that question. Measurements and theory seemed too intractably difficult to be worth the effort. The few people who specialized in aerosols were occupied with practical problems such as their impact on health. Rising public dismay over air pollution drew chemists in to analyze urban smog, which turned out to be a fascinating, and sometimes lethal, mixture of chemical molecules as well as larger particles. Meanwhile, other aerosol experts worked on industrial processes such as clean rooms for manufacturing electronics, and still others investigated military problems such as the way particles scatter laser light. Each researcher had only occasional contact with colleagues who studied other aerosol problems, still less with anyone in other fields of science that might relate to climate.

One respected climatologist who did attack the question was J. Murray Mitchell Jr., inspired by a new kind of aerosol. Studies of fallout from nuclear bomb tests showed that fine dust injected into the stratosphere would linger there for a few years, but it did not cross from one hemisphere to the other. With that in mind, Mitchell pored over global temperature statistics and put them alongside the record of notable volcanic eruptions. In 1961 he announced that large eruptions caused a significant part of the irregular annual variations in average temperature in a given hemisphere. But he could not find a connection between volcanic eruptions and longer-term climate trends. In particular, the warming of the globe in the first half of the century did not match any slackening of eruptions.

In January 1961, on a snowy and unusually cold day in New York City, Mitchell told a meeting of meteorologists that although global temperatures had indeed risen until about 1940, more recently temperatures had been falling. There was so much random variation from place to place and from year to year that nobody had seen the reversal in the millions of weather measurements. Indeed, nobody had made a convincing calculation of average temperatures that covered a large part of the globe. But now Mitchell had painstakingly sifted through the data and ground out a plausible calculation, which showed a clear downturn. None of the available theories of climate change seemed able to explain the recent cooling; Mitchell could only call it "a curious enigma."[2] (See Figure 2, p. 117.)

Through the 1960s meteorologists continued to report relatively cool average global temperatures. Even experts tend to give special attention to the weather that they see when they walk out their doors, and what they saw made them doubt that greenhouse-effect warming was on the way. Yet they had begun to question the comfortable old belief that climate was regulated in a stable natural balance, immune to human intervention.

That sort of view was dwindling almost everywhere. The first Earth Day, held in 1970, marked the emergence of environmentalism into full public awareness and direct political action. To many environmentalists, almost any new technology looked dangerous. Human "interference" with nature began to seem ignorant, reckless, and perhaps wicked. In every democratic industrial nation, citizens pressed their governments to enact environmental protection laws. Governments gave way, taking steps to reduce smog, clean up water supplies, and so forth.

Like everyone else, scientists were affected by the new attitudes. They found it increasingly easy to expect that human activities could warp entire geophysical systems. Reciprocally, the scientists' new concepts and findings were essential in persuading the public that serious environmental harm was at hand. Contemplating the relationship between science and society, some people would say that public opinion responded intelligently to new scientific facts. Others would say that the judgment of scientists bent under the pressure of the mass prejudices of the day. Both views go too far in separating scientific and popular thought. The views of scientists and public evolved together.

A good example of the coevolution of scientific and public views began with Roberts's tentative observation, well in advance of solid scientific inquiry, that jet aircraft might increase cloudiness. The *New York Times* made it a front-page story (23 September 1963), telling its readers that jets "might be altering the climate subtly along major air routes." Such occasional speculations became more serious in the mid-1960s, when the U.S. government announced plans to build a fleet of supersonic transport airplanes. Hundreds of flights a year would inject water vapor and other exhaust into the high, thin stratosphere, where natural aerosols were rare and any new chemical might linger for years. Public opposition grew swiftly. The chief objections (which drove Congress to cancel the project in 1971) were complaints that the fleet would be intolerably noisy and waste taxpayers' money. But the opposition was strengthened by scientists' warnings that emissions from a supersonic fleet could damage the atmosphere.

In 1970 the worries about a supersonic airplane fleet and a number of other environmental issues inspired some policy entrepreneurs to organize the groundbreaking "Study of Critical Environmental Problems." Meeting at the Massachusetts Institute of Technology, some forty experts in a variety of fields deliberated for a month over pollution of the air and oceans, the advance of deserts, and diverse other harms caused by humans. Among other things, they reported that, just as Roberts had suspected, cirrus clouds had increased over the United States in the preceding few decades. Adding a fleet of supersonic airplanes might alter the stratosphere as severely as a volcanic eruption. Or maybe not: the experts admitted that a calculation of actual effects was beyond their reach. In their concluding conference report, as the first item in a list of potential problems the scientists pointed to the global rise of CO_2. Here, too, effects were beyond their power to calculate, and they could only vaguely warn about "widespread droughts, changes of the ocean level, and so forth." The study concluded that the risk of greenhouse warming was "so serious that much more must be learned about future trends of climate change."[3] It was the old heartfelt maxim, More Money Should Be Spent on Research. Nobody thought of recommending actual action to restrict emissions of CO_2.

All but one of the participants in the study were residents of the United States. After all, since 1945 not only climate studies but most fields of science had been dominated by Americans, while every other industrial nation dug out from the devastation of war. But by now the rest of the world had recovered, and it was time for the multinational approach that climate problems required. An assortment of private and government sources funded a comprehensive gathering of experts from fourteen nations, held in Stockholm in

1971. This "Study of Man's Impact on Climate" was the first major conference to focus entirely on that subject. Exhaustive discussions brought no consensus on what was likely to happen, but all agreed that serious changes were possible. For example, we might pass a point of no return if the Arctic Ocean's ice cover disappeared—an illustration of "the sensitivity of a complex and perhaps unstable system that man might significantly alter." The widely read report concluded with a ringing call for attention to humanity's emissions of particle pollutants and greenhouse gases. The climate could shift dangerously "in the next hundred years," the scientists declared, "as a result of man's activities."[4]

Weather disasters in the news reinforced the somber new mood. In 1972 a drought ravaged crops in the Soviet Union and the Indian monsoon failed. In the United States the Midwest was struck by droughts severe enough to show up repeatedly on the front pages of newspapers and on television news programs. World food prices soared to an all-time high. Most dramatic of all, years of drought in the African Sahel reached an appalling peak: millions were starving, hundreds of thousands died, and mass migrations disrupted entire nations. Television and magazine pictures of sun-blasted fields and emaciated refugees brought home just what climate change could signify for all of us.

It was not only the public mood that prepared scientists to think about rapid and destructive climate change. They were also impressed by new data about past climates. Some of the strongest came from the University of Wisconsin, where Bryson had gathered a group to take a new, interdisciplinary look at climate. The group included an anthropologist who studied the native American cultures of the Midwest. From bones and pollen the scientists deduced that a disastrous drought had struck the region in the 1200s—the very period when the civilization of the Mound Builders had collapsed. It was already known that around that time a great drought had ravaged the Anasazi culture in the Southwest (the evidence was constricted tree rings in ancient logs from their dwellings). Compared to the drought of the 1200s, the ruinous Dust Bowl of the 1930s had been mild and temporary. A variety of historical evidence hinted that the climate shift had been worldwide. And there seemed to have been distinct starting and ending points. By the mid-1960s, Bryson concluded that "climatic changes do not come about by slow, gradual change, but rather by apparently discrete 'jumps' from one [atmospheric] circulation regime to another."[5]

Next, Bryson's group reviewed radiocarbon dates of pollen from around the end of the last ice age. In 1968 they reported a rapid shift in the mix of tree species that occurred around 10,500 years ago. Up to this point, when climate scientists spoke of "rapid" change they had meant something happening in as little as a thousand years. Bryson and his collaborators saw changes within a century. Looking at hundreds of radiocarbon dates spanning the past dozen millennia, the group arrived at a disturbing general conclusion. Periods of "quasi-stable" climate ended in catastrophic "discontinuities" when "dramatic climate change occurred in a century or two at most."[6]

To be sure, it did not require global climate change to transform some particular forest—strictly local events could do that. Many experts continued to feel it was pure speculation to imagine that the entire world's climate could change in less than a thousand years or so. If Bryson and some other climate scientists were now willing to announce that their data showed catastrophes, that was partly because the general mood of the times allowed it. At the same time, such announcements, picked up and disseminated by science journalists, were helping to shift the general mood. Scarcely any popular article on climate in the 1970s lacked a Bryson quote or at least a mention of his ideas.

Bryson's ideas could not be brushed aside, for corroborating evidence was showing up from the ends of the Earth. The first reports came from the frozen heights of Greenland, which had already played an important role in the nineteenth-century controversy over the ice ages. After a few geologists had dared to postulate the existence, in the distant past, of kilometer-thick masses of ice, astonished explorers of Greenland had found just such a thing beneath their skis. Built up layer by layer over many centuries, the ice sheet carried frozen within it a record of the past.

The first attempts to exploit the ice record had begun during the 1957–1958 International Geophysical Year. A few scientists had traveled to Camp Century, a military installation high on the Greenland ice cap. The daunting logistics were handled by the U.S. government, eager to master the Arctic regions that lay on the shortest air routes to the Soviet Union. In 1961 a specially adapted drill brought up cores of ice 5 inches in diameter in segments several feet long. This was no small feat in a land where removing your gloves for a few minutes to adjust something might cost you the skin on your fingertips, if not entire fingers.

After this proof of feasibility, scientists dreamed of penetrating ever deeper into the ice, and thus into the past. Ice drillers began to form their own little specialty, crafting international collaborations to venture onto the world's ice sheets. The divergent interests of people from different nations made for long and painful negotiations. But the trouble of cooperating was worth it to bring in a variety of expertise—and a variety of agencies that might grant funds. Devising ingenious equipment, scientists managed to transport it by the ton to the ends of the earth. Costly drill heads might get irretrievably stuck half a mile down, but engineers went back to their drawing boards, team leaders contrived to get more funds, and the work slowly pushed on.[7] After five years of effort, the Camp Century drill reached bedrock at a depth of some 1.4 kilometers (7/8 of a mile), bringing up ice as much as 100,000 years old. Two years later, in 1968, another team retrieved a long core of ancient ice from a site even colder and more remote, Antarctica.

Much could be read from the cores. For example, individual layers with a lot of acidic dust pointed to past volcanic eruptions. Larger amounts of mineral dust turned up deep in the ice, evidence that during the last ice age the world climate had been windier: storms had carried dust to Greenland clear from China. But the greatest hopes centered on bubbles in the ice. By good fortune there was this one thing on the planet that preserved ancient air intact, a million tiny time capsules. For a long time, however, nobody could figure out how to extract and measure the air reliably.

In the early years the most useful work was done with the ice itself, using a method conceived back in 1954 by an ingenious Danish glaciologist, Willi Dansgaard. He showed that the ratio of oxygen isotopes (O^{18}/O^{16}) in the ice measured the temperature of the clouds at the time the snow had fallen—the warmer the air, the more of the heavy isotope got into the ice crystals. It was an exhilarating day for the researchers at Camp Century, who had been making measurements along each cylinder of ice as it was pulled up from the borehole, when they saw the isotope ratios change and realized they had reached the last ice age. The preliminary study of the ice cores, published in 1969, showed variations that indicated changes of perhaps 10°C. Comparison of Greenland and Antarctic cores showed the climate changes were truly global, coming at essentially the same time in both hemispheres. That immediately tossed into the wastebasket some old theories of the ice ages that relied on merely regional circumstances.

Study of ice cores confirmed a detail that Wally Broecker had already noticed in deep-sea sediments: the glacial cycle followed a sawtooth curve. In each cycle a spurt of rapid warming was followed by a more gradual, irregular descent back into the cold over tens of thousands of years. Warm intervals, like the climate through the history of human civilization, normally did not last long on this geological timescale. Beyond such fascinating hints, however, the Greenland ice cores could say little about long-term cycles. For the ice at great depths flowed like tar, confusing the record. Despite the arduous efforts of the ice drillers, in the 1970s the most reliable data were still coming from deep-sea cores.

That work, manhandling long, wet pipes on a heaving deck, was also strenuous and hazardous. Oceanographers, like ice drillers, lived close together for weeks or months at a time under spartan conditions, far from their families. The teams might function smoothly—or not. Either way, the scientists labored long hours, for the problems were stimulating, the results could be exciting, and dedication to work was the norm.

To make it all worthwhile, the ocean-bed drillers had to draw on all their knowledge and luck to find the right places to sample. In these few places, layers of silt had built up on the ocean bed unusually swiftly and steadily, and without subsequent disturbance. During the 1960s engineers worked out drilling techniques to extract continuous 100-meter cores of clay (what Cesare Emiliani had been asking for since the 1950s). Meanwhile, scientists devised clever methods for extracting data that could shed light on past climates.

The most prominent feature was a 100,000-year cycle, so strong that it had to be a key to the entire climate puzzle. This long-term cycle had been tentatively identified in several earlier studies of sea-bed cores. Corroboration came from a wholly different type of record. In a brick-clay quarry in Czechoslovakia, George Kukla noticed how windblown dust had built up deep layers of soil (what geologists call "loess"). The multiple advances and retreats of ice sheets were visible to the naked eye in the colored bands of different types of loess. It is one of the few cases in this story where traditional field geology, tramping around with your eyes open, paid a big dividend. Analysis showed that here, too, at the opposite end of the world from some of the deep-sea drilling sites, the 100,000-year cycle stood out.

But nobody could be entirely sure; radiocarbon decayed too rapidly to give dates going back more than a few tens of thousands of years. A deeper timescale could be estimated only by measuring lengths down a core, and it was uncertain whether the sediments were laid down at a uniform rate. In 1973 Nicholas (Nick) Shackleton nailed down the dates with a new technique that used radioactive potassium, which decays much more slowly than radiocarbon. He applied the timescale to a splendid deep-sea core pulled from the Indian Ocean, the famous core *Vema* 28-238 (named after Lamont's oceanographic research vessel). This core reached back over a million years, carrying a huge stretch of precise data, and Shackleton analyzed it ingeniously. When he showed his graph to a roomful of climate scientists, a spontaneous cheer went up.

Shackleton's graph, confirmed by other long cores, at last proved beyond doubt what Emiliani had stoutly maintained: there had been not four major ice ages, but dozens. A sophisticated numerical analysis of the data found an entire set of favored frequencies. The ice sheets had waxed and waned in a complex rhythm set by cycles with lengths roughly 20,000 and 40,000 years long, as well as the dominant 100,000 years. These numbers were in approximate agreement with types of orbital variations in Milankovitch's original calculations. Particularly impressive was a high-precision 1976 study of Indian Ocean cores that split the 20,000-year cycle into a close pair of cycles with lengths of 19,000 and 23,000 years. That was exactly what the best new astronomical calculations predicted for the "precession of the equinoxes," a wobbling of the Earth's axis. These studies left most scientists convinced that orbital variations were central to long-term climate change.

There remained the old objection that the subtle changes in the amount of sunlight reaching the Earth were far too small to affect climate. Worse, the record was dominated by the 100,000-year cycle, which brought particularly tiny variations of sunlight (caused by a minor change in eccentricity, a slight stretching of the Earth's path around the sun away from a perfect circle). The only reasonable explanation, as Shackleton and others immediately understood, must be that feedbacks were amplifying the changes. Presumably the feedbacks involved other natural cycles that resonated at roughly the same timescale. The orbital variations served only as a "pacemaker" that pinned down the exact timing of internally driven feedback cycles. In Milankovitch's day most climate scientists had thought such a thing unlikely. By the mid-1970s scientists were seeing feedback cycles everywhere, poised to react with hair-trigger sensitivity to external influences.

What were the natural cycles that fell into step with the shifts of sunlight? The most obvious suspect was the continental ice sheets. It took many thousands of years for snowfall to build up until the ice began to flow outward. A related suspect was the solid crust of the Earth. On a geological scale it was not truly solid; it sluggishly sagged where the great masses of ice weighed it down, and sluggishly rebounded when the ice melted. (Scandinavia, relieved of its icy burden some 20,000 years ago, is still rising a few millimeters a year.) Since the 1950s scientists had speculated that the timing of glacial periods might be set by these slow plastic flows, the spreading of ice and the warping of crustal rock. During the 1970s a number of scientists invented elaborate numerical models that suggested how 100,000-year cycles might be driven by feedbacks among ice buildup and flow, along with the associated movements of the Earth's crust, changes in reflection of sunlight, and the rise and fall of sea level. The models were obviously speculative, for nobody had reliable equations or data for such processes. But it did seem possible that some kind of natural feedback systems could amplify the weak Milankovitch sunlight changes (and perhaps other variations too?) into full-blown ice ages.

Whatever caused it, the curve of ancient temperatures followed the complex pattern of astronomical cycles with uncanny accuracy. It was natural to continue the calculations and extrapolate the curve forward in time. The predicted curve headed downward for the next 20,000 years or so. Emiliani, Kukla, Shackleton, and other specialists concluded that the Earth was gradually heading into a new ice age.

Or perhaps not so gradually, on a geological timescale. In 1972 Murray Mitchell, the respected climate expert from the U.S. Weather Bureau, said that in place of the old view of "a grand, rhythmic cycle," new evidence was revealing a "much more rapid and irregular succession" in which the Earth "can swing between glacial and interglacial conditions in a surprisingly short span of millennia (some would say centuries)."[8] Evidence for these swings, in particular the Younger Dryas oscillation around 12,000 years ago, had turned up in radiocarbon studies of ancient glacier moraines, fossil shorelines, lake levels, shells in the most undisturbed deep-sea cores, and more. The most convincing evidence came from the Greenland core drilled by Dansgaard's group of Danes and Americans. Mixed in with the gradual cycles were what Dansgaard called "spectacular" shifts, lasting perhaps as little as a century or two—including, once again, the Younger Dryas.[9] Perhaps that was an illusion, for the flow of ice at great depths muddled the record. Or perhaps back then the climate around the North Atlantic, if not around the entire globe, really had changed drastically.

If scientists were increasingly willing to consider that global climate might change greatly in the space of a century or so, it was because different pieces of the puzzle seemed to fit together better and better. Sudden lurches had persistently turned up in the rudimentary weather and climate models that were built in the 1950s and 1960s from rotating dishpans or from sets of simple equations run through computers. Scientists could have dismissed these models as too crude to say anything reliable—but the historical data showed that the notion of radical climate instability was not absurd after all. And the researchers could have dismissed the jumps in the data as artifacts due to merely regional changes or simple errors—but the models showed that such jumps were physically plausible.

Equations that scientists were developing to model continental ice sheets offered disturbing prospects of abrupt changes. In 1962 John Hollin had argued that the great volumes of ice in Antarctica, piled up kilometers high and pushing slowly toward the ocean, were held in place by their fringes, pinned at the "grounding line," where they rested on the ocean floor. A rise in sea level could float an ice sheet up off the seafloor, releasing the entire stupendous mass behind it to flow more rapidly into the sea. The idea intrigued Alex Wilson, who pointed to the spectacle of a "surge." Glaciologists had long known that a mountain glacier might suddenly give up its usual slow creeping to race forward at hundreds of meters a day. They figured this happened when the pressure at the bottom melted ice so that water lubricated the flow. As the ice began to move, friction melted more water and the flow accelerated. Could the ice in Antarctica become unstable in this fashion?

If so, the consequences sketched by Wilson would be appalling. As the ice surged into the sea, the world's seacoasts would flood. But that would be almost the least of humanity's problems. Immense sheets of ice would float across the southern oceans, cooling the world by reflecting sunlight, bringing on a new ice age.

Through the 1960s few scientists gave much credence to these ideas. The ice that covered most of Antarctica, in places over four kilometers thick, seemed firmly grounded on the continent's bedrock. But around 1970 John Mercer, a bold and eccentric glaciologist at Ohio State University, drew attention to the West Antarctic Ice Sheet. This is a smaller (but still enormous) mass of ice, separated by a mountain range from the bulk of the continent. Mercer argued that this mass was held back in an especially delicate balance by the ice shelves floating at its rim. The shelves might disintegrate under a slight warming. If the West Antarctic Ice Sheet was released and slid into the oceans, the sea level would rise as much as 5 meters (16 feet), forcing the abandonment of many great cities. This could happen rapidly, Mercer thought, conceivably within the next forty years.

Other glaciologists found this interesting enough to prompt them to build simple models of ice movements, using data from adventurous survey expeditions that had traversed parts of Antarctica during the International Geophysical Year and after. They found that the West Antarctic Ice Sheet might indeed be unstable. It was "entirely possible," one author concluded in 1974, that the sheet was already now starting to surge forward.[10] Everyone admitted that these models were highly speculative. It seemed physically almost impossible that the West Antarctic sheet could disintegrate in less than a few centuries at worst. But a surge that dumped a fifth of a continent of ice into the oceans over the next few centuries would be no small thing, and the models were too simplified to entirely rule out an even faster collapse.

If the climate experts of the 1970s seem to have been a bit preoccupied with ice, that fitted with what was on the public's mind. For science journalists liked to write about possible future deluges and ice ages, spectacles far better at catching a reader's eye than theories of a gradual warming. As for the climate experts themselves, for a century their profession had concerned itself above all with ice ages. Their field studies were devoted to measuring the swings between warm and glacial epochs. Home at their desks, they attacked the grand challenge of explaining what might cause these swings. Now that they were beginning to turn their attention from the past to the future, the most natural meaning to attach to "climate change" was the next swing into cold.

In 1972 a group of leading glacial-epoch experts met at Brown University to discuss how and when the present warm interglacial period might end. Reviewing the Greenland ice cores, Emiliani's foraminifera, and other field evidence, they agreed that interglacial periods tended to be short and to end relatively quickly. A large majority further agreed that extrapolating the Milankovitch curves into the future showed that "the natural end of our warm epoch is undoubtedly near." They noted that weather records showed the world had not been warming since 1940. The scientists were far from reaching agreement; some insisted that any cooling could be counteracted by greenhouse-effect warming, or by other, unknown factors. But the majority agreed that serious cooling "must be expected within the next few millennia or even centuries."[11] Several members of the study group wrote a letter to President Richard Nixon, calling on the government to support intensified climate studies. That was one example of a general movement during the 1970s. Scientifically trained people were making contact with policy elites to address the planet's environmental future.

The study group suspected that natural cycles were not the greatest hazard. While some worried about CO_2 warming the planet, others now saw a greater risk of cooling. Would the drop in temperature that the Milankovitch schedule predicted in future millennia be accelerated by smoke and dust from human agriculture and industry? Bryson, for one, argued with increasing conviction that such a cooling was only too likely.

As Bryson had noticed while flying over the tropics, entire regions could be hazed over for months at a time. Already in 1958, one expert had remarked that "there can no longer be any sharp division between polluted and unpolluted atmospheres."[12] Most meteorologists, however, were slow to notice the spread of pollution beyond cities. Better understanding came only after people studying smog set up a network of stations that regularly monitored the atmosphere's turbidity (haziness). In 1967 two scientists at the National Center for Air Pollution Control in Cincinnati reported a gradual increase in the general turbidity over regions spanning a thousand kilometers. Further checks of the record of turbidity turned up increases even in remote areas like Hawaii and the North and South Poles. Could humanity's emissions be affecting the global climate not in some abstract future, but right now? The world's oceans and air could no longer be seen as a virtually infinite dumping ground that could safely absorb any and all emissions. Concern grew following reports that the air over the North Atlantic was twice as dirty in the late 1960s as it had been in the 1910s, which suggested that the natural processes that washed aerosols out of the atmosphere were not keeping up with human emissions.

But how much of the haze was really caused by humans? In 1969 Mitchell pushed ahead with his statistical studies of temperatures and volcanoes. He calculated that about two-thirds of the cooling seen in the Northern Hemisphere since 1940 was due to a few volcanic eruptions. He concluded that "man has been playing a very poor second fiddle to nature as a dust factory."[13] Other climatologists could not agree on how strong the effect of volcanic eruptions was, and how strong the effect of human pollution. They could only admit that these questions had been overlooked for too long and deserved sustained scrutiny.

In 1971 S. Ichtiaque Rasool and Stephen Schneider entered the discussion with a pioneering numerical computation. (This was the first atmospheric science paper by Schneider, who would later become a well-known commentator on global warming.) Developing ideas suggested by Mitchell and others, Rasool and Schneider explained how the effect of aerosols could vary. Some kinds of haze might not cool the atmosphere after all, but warm it. It depended on how much the aerosols absorbed radiation coming down from the Sun, and how much they trapped heat radiation rising up from the Earth's surface. Rasool and Schneider's calculation gave cooling as the most likely result. If the pollution continued to rise rapidly, they exclaimed, it "could be sufficient to trigger an ice age!"[14] In fact, their equations and data were oversimplified. At the frontier of research, nobody gives much credence to any publication until it has withstood inspection and criticism, and scientists (including Schneider himself) soon pointed out crippling flaws. But if this paper was wrong, what was the correct interpretation?

Beyond the direct effects of aerosols as they absorbed or scattered radiation, there remained an even greater mystery: how did particles help create particular types of clouds? And beyond that loomed an equally enigmatic problem: how did a given type of cloud reflect radiation coming from above or below, to cool the Earth or perhaps warm it? Simplified calculations turned up all sorts of subtle and complex influences. The one sure thing was that aerosols could make a difference to climate.

Many people now wanted to know just how big a difference aerosols or CO_2 or other human products could make. To answer that grave question, scientists needed something better than crude, hand-waving models. They turned to mathematical calculations. In 1963 Fritz Möller, building on the pioneering work of Gilbert Plass (Chapter 2), had produced a model for how radiation moves up and down a typical column of air. His key assumption was that the humidity of the entire atmosphere should increase with increasing temperature. To put this into the calculations, he held the relative humidity constant, which was just what Arrhenius had done in his pioneering calculations back in 1896 (Chapter 1). Möller got the same amplifying feedback that Arrhenius had found. As the temperature rose, more water vapor would remain in the air, adding its share to the greenhouse effect.

When he finished his calculation, Möller was astounded by the result. Under some reasonable assumptions, doubling CO_2 in the atmosphere could bring a temperature rise of 10°C—or perhaps even more, for the mathematics would allow an arbitrarily high rise. More and more water would evaporate from the oceans until the atmosphere filled with steam! Möller found this result so implausible that he doubted the whole theory. Indeed, his method was later shown to be fatally flawed. But most research begins with flawed theories, which prompt people to propose better ones. Some scientists found Möller's calculation fascinating. Was the mathematics trying to tell us something truly important? It was a disturbing discovery that a simple calculation (whatever problems it might have in detail) could produce a catastrophic outcome. That was one stimulus for taking up the challenging job of building full-scale computer models.

Through the 1960s, however, computer simulations of the entire general circulation of the atmosphere remained primitive. Even if the calculations ran for weeks on end, the results looked only roughly like the present climate. That could scarcely tell people how things might change under a subtle influence. So some climate scientists tried using computers in a less expensive and arduous way. They built highly simplified models that could work out rough numbers in a few minutes. Although some scientists gave these simple models no credence, others felt that they were valuable "educational toys"—a helpful starting point for testing assumptions and for identifying spots where future work could be fruitful.[15]

The most important simple model of climate change was built by Mikhail Budyko in Leningrad. He was drawn to the issue by grandiose proposals that had concerned Soviet climatologists since the 1950s. How about diverting rivers from Siberia, for example, to turn southern deserts into farmland? It sounded good at first, but what would it do to the climate of Siberia itself? Or how about warming the Arctic by spreading soot over snow and ice to absorb sunlight? Budyko studied historical data on how the snow cover in a region connected with temperature, and he found a dramatic interdependence.

To pin down his ideas, around 1968 Budyko constructed a highly simplified set of equations. They represented the heat balance of the Earth as a whole, summing up over all latitudes the incoming and outgoing radiation: an "energy budget" model. When he plugged plausible numbers into his equations, he found that for a planet under given conditions—that is, a particular atmosphere and a particular amount of radiation from the Sun—more than one state of glaciation was possible. If the planet had arrived at the present state after cooling down from a warmer climate, the reflection of sunlight ("albedo") back into space by the dark sea and soil would be relatively low. So the planet could remain entirely free of ice. If it had come to the present by warming up from an ice age, keeping some snow and ice that reflected sunlight, it could retain its chilly ice caps.

Under present conditions, the Earth's climate looked stable in Budyko's model. But not too far above present temperatures and albedo, the equations reached a critical point. The global temperature would shoot up as the ice melted away entirely, uncovering sunlight-absorbing soil and water. That would result in a uniformly and enduringly warm planet with high ocean levels, which it was in the time of the dinosaurs. And if the temperature dropped not too far below present conditions, the equations hit another critical point. Here temperature could drop precipitously as more and more water froze, until the Earth reached a stable state of total glaciation: the oceans entirely frozen over, the Earth transformed permanently into a gleaming ball of ice! Budyko thought it possible that our era was one of "coming climatic catastrophe . . . higher forms of organic life on our planet may be exterminated."[16]

Other analysts were on the same trail, independently of Budyko's work in Leningrad; communications were sporadic across the Cold War frontiers. What finally caught full attention was an energy-budget model published in 1969 by William Sellers at the University of Arizona. His model was still "relatively crude," as Sellers admitted (adding that this was unfortunately "true of all present models"), but it was straightforward and elegant. Climatologists were impressed to see that although Sellers used equations quite different from Budyko's, his model too could approximately reproduce the present climate—and it too showed a cataclysmic sensitivity to small changes. If the energy received from the Sun declined by 2 percent or so, whether because of solar variations or increased dust in the atmosphere, Sellers thought another ice age could result. Beyond that, Budyko's nightmare of a totally ice-covered Earth seemed truly possible. At the other extreme, Sellers suggested, "man's increasing industrial activities may eventually lead to a global climate much warmer than today."[17]

Did the Budyko-Sellers catastrophes reflect real properties of the global climate system? That was a matter of brisk debate. In the early 1970s some scientists did find it plausible that feedbacks could build up a continental ice sheet more rapidly than had been supposed. The opposite extreme—a self-sustaining heating—might be even more catastrophic. For there was new evidence that another calculation from a few equations, Möller's preposterous runaway greenhouse, represented something real.

A planet is not a lump in the laboratory that scientists can subject to different pressures and radiations, comparing how it reacts to this or that. We have only one Earth, and that makes climate science difficult. To be sure, we can learn a lot by studying how past climates were different from the present one. But these are limited comparisons—different breeds of cat, but still cats. Fortunately, our solar system contains wholly other species, planets with radically different atmospheres.

Radio observations of the planet Venus, published in 1958, indicated that the surface temperature was amazingly hot, around the melting point of lead. Why would a planet about the same size as the Earth, and not all that much closer to the Sun, be so drastically different? In 1960 a young doctoral student, Carl Sagan, took up the problem and got a solution that made his name known among astronomers. Using what he later recalled were embarrassingly crude methods, taking data from tables designed for steam-boiler engineering, he showed that the greenhouse effect could make Venus a furnace. He thought this was due mainly to water vapor, in a self-perpetuating process. The surface had been so warm that whatever water the planet possessed had gone into the atmosphere as vapor, where it helped maintain the extreme greenhouse-effect condition. It was later found that Venus's atmosphere has little water (it is mostly CO_2). Sagan was wrong—another of those scientific mistakes that usefully stimulated further work.

And then there was Mars. In 1971 the spacecraft *Mariner 9,* a marvelous jewel of engineering, settled into orbit around Mars and saw . . . nothing. A great dust storm was shrouding the entire planet. Such storms are rare for Mars, and this one was no misfortune for the observers, but great good luck. They immediately saw that the dust had profoundly altered the Martian climate, absorbing sunlight and heating the planet by tens of degrees. The dust settled after a few months, but its lesson was clear. Haze could warm an atmosphere. More generally, anyone studying the climate of *any* planet would have to take dust very seriously. Moreover, it seemed that the temporary warming had reinforced a pattern of winds that had kept the dust stirred up. It was a striking demonstration that feedbacks in a planet's atmospheric system could flip weather patterns into a drastically different state. That was no longer speculation but an actual event in full view of scientists— "the only global climatic change whose cause is known that man has ever scientifically observed."[18]

Before *Mariner* arrived at Mars, Sagan had made a bold prediction. He suggested that the Red Planet's atmosphere could settle in either of two stable climate states. Besides the current age of frigid desiccation, there was another possible state, more clement, which might even support life. The prediction was validated by crisp images of the surface that *Mariner* beamed home after the dust cleared. Geologists saw evidence that vast water floods had ripped the planet in the distant past. Calculations suggested the planet's climate system could have been flipped from one state to another by relatively minor changes.

Such theories of change now dominated thinking about Earth's climate, too. People no longer imagined the climate was permanently stable. It was not the weight of any single piece of evidence that convinced them, but the accumulation of evidence from different, independent fields. Mars and Venus, increased haze and jet contrail clouds, ancient catastrophic droughts, fluctuating layers in ice and in seabed clays, computer calculations of planetary orbits and of energy budgets and of ice sheet collapse: each told a story of climate systems prone to terrible lurches. Each story, bizarre in itself, was made plausible by the others. Most experts knew only some parts of the evidence. Most members of the public knew hardly any of it. But the main idea got around.

To be sure, no tremendous climate change had come in recent memory. But people were ready to take a longer view. During the first half of the century, violent interruptions of life by wars and economic upheavals had made long-term planning seem pointless. By the 1970s more settled times encouraged people to think farther ahead. If human emissions were liable to change the climate in the twenty-first century, that no longer seemed too far away to plan for.

Maybe somebody should do something about it?

CHAPTER FIVE

PUBLIC WARNINGS

The politics of science came up briefly at a 1972 symposium where scientists were discussing the rising level of CO_2 in the atmosphere. Should they make a statement calling for some kind of action? "I guess I am rather conservative," one expert remarked. "I really would like to see a better integration of knowledge and better data before I would personally be willing to play a role in saying something political about this." A colleague replied: "To do nothing when the situation is changing very rapidly is not a conservative thing to do."[1]

Most scientists felt they were already doing their jobs by pursuing research and publishing it. Anything important would presumably be noticed by science journalists and government science agencies. For really important problems the scientists could convene a study group, perhaps under the National Academy of Sciences, and issue a report. Experts like Roger Revelle and Reid Bryson were more than willing to explain their ideas when asked, and they might even make an effort to come up with quotable phrases for reporters. They were glad to give talks on the state of climate science, or write an article for a magazine like *Scientific American*. Such efforts would reach, if not exactly the public, the small segment of the public that was well educated and interested in science.

A wider public would take notice only if something special came along, something newsworthy. In the early 1970s the climate offered plenty of such opportunities. The dramatic crop failures in India, Russia, Africa, and the American Midwest sent journalists to talk with climate scientists time and again. They reported that some scientists suspected the weather fluctuations could be the harbinger of the next ice age. After all, many scientists now accepted that the long-term trend, extrapolating the Milankovitch curves, should be a cooling. The journalists did not always report that nearly all these scientists expected the cold would not become severe for centuries, or more likely millennia. And if that was what would naturally happen, what might human intervention do? The expert most quoted on that was Bryson. He wanted everyone to know about the increase in smoke and dust caused by industry, deforestation, and overgrazing. Like the smoke from a huge volcanic eruption, he said, the "human volcano" could bring a disastrous climate shift. The effects, he declared, "are already showing up in rather drastic ways." As rising population crashed against the increasingly erratic weather, the world might face mass starvation.[2]

Climate scientists insisted that such ideas were by no means scientific predictions, but only possibilities. (Thirty years later they were still debating whether the rise of pollution had in fact caused the great Sahel drought.) Responsible journalists made it clear that every expert, including Bryson, admitted ignorance. Yet the majority of scientists did feel, as *Time* magazine put it, that "the world's prolonged streak of exceptionally good climate has probably come to an end—meaning that mankind will find it harder to grow food."[3] The most common scientific viewpoint was summed up by a scientist who explained that the rise in dust pollution worked in the opposite direction from the rise in CO_2, so nobody could say whether there would be cooling or warming. The real point was a broader one: "We are entering an era when man's effects on his climate will become dominant."[4]

Whether our emissions would speed the coming of a new ice age or bring on greenhouse warming, the moral lesson seemed the same. "We have broken into the places where natural energy is stored and stolen it for our own greedy desires," a journalist declaimed. Some people, he said, expect that "our tampering with the delicate balances of nature" would call down "the just hand of divine judgment and retribution against materialist sinners."[5]

More optimistic people suggested that we might counter any ill effects by deliberately bending the climate to our will. If an ice age approached, we might spread soot from cargo aircraft to darken the Arctic snows. If warming threatened, we might inject sunlight-reflecting smog into the stratosphere. Most scientists dismissed such ideas, but not because they sounded like science fiction. It seemed only too plausible that humanity could alter the climate. But the bitter fighting between communities over cloud seeding would be as nothing compared with the conflicts that attempts to engineer global climate might bring. What nation would be trusted to alter other nations' weather? And our knowledge was so primitive that even the best-meant intervention might only make things worse.

The general public was aware of all this talk, but only vaguely. Since the nineteenth century, news media had trumpeted future threats from this or that quarter, and the public saw nothing special here. Reports of scientific findings were usually relegated to a few paragraphs on the inside pages of the better newspapers or in the science-and-culture section of newsmagazines, which reached only the more alert citizens. It was Nigel Calder, a well-known British science journalist, who first reached a somewhat broader public— those who watched educational shows on public television—to the threat of climate catastrophe. In 1974 he produced a two-hour feature on weather, which spent a few minutes warning of a possible "snowblitz." Entire countries could be obliterated under layers of snow; billions would starve. The new ice age "could in principle start next summer, or at any rate during the next hundred years."[6]

Most experts despised such talk. They felt that the public was being led astray by a small number of scientists and sensation-seeking journalists. The official message from the director-general of the United Kingdom Meteorological Office was: "No need for panic induced by the prophets of doom." Like many meteorologists, he held that "the climatic system is so robust . . . that man has still a long way to go before his influence becomes great enough to cause serious disruption."[7] The traditional belief in a benign Balance of Nature was still widely held among scientists as well as the public.

There were only a few scientists like Bryson who took the prospect of climate catastrophe so seriously that they made a personal effort to address the public directly. The climatologist who worked hardest to make his ideas heard was Stephen Schneider. He and his wife, a journalist, wrote a popularizing book, *The Genesis Strategy: Climate and Global Survival.* Insisting that climate could change more quickly and drastically than most people imagined, they advised the world to devise policies to cushion the shocks, for example by building a more robust agricultural system. As Joseph had advised Pharaoh in the Book of Genesis, we should prepare for lean years to follow fat ones.[8]

Some scientists criticized Bryson, Schneider, and others for speaking directly to the public. The time spent writing a book and going about the country delivering public lectures was time away from doing "real" science. Besides, the whole subject was so riddled with uncertainties that it seemed unfit for presentation, in a few simplified sound bites, to the scientifically naive public.

But like it or not, the issue was becoming political, at least in the narrow sense that policies were at stake. At professional meteorological conferences, debates over technical questions such as the rate of CO_2 buildup became entangled with debates over how governments should respond. Scientists began to struggle with questions far beyond their professional expertise. Should reliance on fossil fuels be reduced? How much money should be spent on averting climate change, when so much was needed for the struggle to feed the world's poor? And was it even proper for a scientist to speak, as a scientist, on such questions?

Unable to agree even whether the world was likely to get warmer or colder, the scientists did unanimously agree that the first step must be to redouble the effort to understand how the climate system worked. Calls for research always came naturally to researchers, but from the early 1970s onward climate scientists issued such calls with increased frequency and passion. They had never seen such strong reasons to insist on their traditional principle, More Money Should Be Spent on Research.

The science of long-term climate change nevertheless remained a minor topic. A rapid rise in publications had begun around 1950, but that did not mean much, for the starting level had been negligibly small. Around 1970 the advance had stalled, and the rate of publication was now rising only sluggishly. Into the mid-1970s, well under a hundred scientific papers per year were published worldwide on any aspect of the subject.[9] During those economically stagnant times, the funding for climate science in every country was generally static. One reason was that the funds were dispersed among a variety of private organizations and relatively small and weak government agencies.

One chance for improvement did come in 1970. Funding of the ocean sciences was particularly scattered, and in the United States several groups had been promoting a centralized agency. With a boost from the rise of environmentalism, they prodded the U.S. government not only to consolidate the nation's maritime programs, but to merge them with atmospheric research. The result was a new organization, the National Oceanic and Atmospheric Administration, which was to address the planet's entire fluid envelope. From the outset, NOAA was one of the world's chief sources of funding for basic climate studies. The agency was created, however, by rearranging programs without adding new money. And if

NOAA had a central focus, it was in developing economically important resources such as fisheries. Climate scientists continued to look for an organization and dedicated funding all their own.

In fact, the study of climate had not yet reached a point where a large-scale coordination of research programs was necessary. In each separate specialty, knowledge was so primitive that a group could go far by pursuing its own questions, getting funds wherever it could, and only occasionally swapping notes with other specialists. For the great trick of science is that you don't have to understand everything at once. The climate scientists of the 1970s were not like the policy makers who have to make decisions that take into account all the confusion of business or politics. Scientists could pare down a system into something so simple that they had a chance to fully understand it—that is, if they dedicated their lives to the task. Among those who did this was Syukuro Manabe.

"Suki" Manabe was one of a group of young men who graduated from Tokyo University in the difficult years just after the Second World War and looked for a career in meteorology. Ambitious and independent-minded, they found few opportunities for advancement in Japan and wound up making their careers in the United States. In 1958 Manabe was invited to join the computer modeling group founded by John von Neumann. After the group had achieved its breakthrough in 1955 with a model that produced realistic-looking regional weather (Chapter 3), von Neumann had drummed up government funding for a project with an even more ambitious goal. His team would construct a general circulation model of the entire three-dimensional global atmosphere, stable enough to derive the planet's actual climate system directly from the basic physics equations for fluids and energy. The effort got under way at the U.S. Weather Bureau in Washington, D.C., under the direction of Joseph Smagorinsky. (It later moved to Princeton.)

Smagorinsky's best idea was to recruit Manabe and work with him on the proposed climate model. Starting with winds, rain, snow, and sunlight, they added the greenhouse effect of water vapor and CO_2; they put in the way moisture and heat were exchanged between air and the surfaces of ocean, land, and ice, and much more. Manabe spent many hours in the library researching such esoteric topics as the way various types of soil soak up water. The equations too needed attention: the system had to be efficient to calculate, and not prone to veer off into wildly unrealistic numbers.

By 1965 Manabe and Smagorinsky had a three-dimensional model that solved the basic equations for an atmosphere divided into nine levels. It was highly simplified, with no geography—everything was averaged over zones of latitude, and land and ocean surfaces were blended into a sort of swamp that exchanged moisture with the air but could not take up heat. Nevertheless, the way the model moved water vapor around the planet looked gratifyingly realistic. The printouts showed a stratosphere, a zone of rising air near the equator (creating the doldrums that becalmed mariners), a subtropical band of deserts, and so forth. Still, many of the details came out wrong.

As it became apparent that such modeling could bring useful results, more groups joined the effort. Central to their progress was the headlong advance of electronic computers: from the mid-1950s to the mid-1970s, the power available to modelers increased by a factor of several thousand. Through the 1960s and 1970s, important general circulation model (GCM) groups appeared at institutions from New York to Australia.

Particularly influential was a group at the University of California, Los Angeles, where Yale Mintz had recruited another young Tokyo University graduate, Akio Arakawa, to help with the mathematics. In 1965 Mintz and Arakawa produced a model that, like Smagorinsky and Manabe's, bore some resemblance to the real world. Another important effort was started in 1964 at the National Center for Atmospheric Research (NCAR) in Boulder, Colorado. The leaders were Warren Washington and yet another Tokyo graduate, Akira Kasahara. But the premier model remained Manabe's.

Although modeling had become a large combined effort, it was not yet able to force through the thicket of problems. The best computers of the day, using up weeks of costly running time, could calculate only a very crude simulation of a typical year of global climate. The workers figured they would need a hundred times more computer power to do the job right. They would get that within another decade or two. Yet even if the computers had been a million times faster, the simulations would have been unreliable. For the modelers were still facing that famous limitation of computers, "garbage in, garbage out."

The calculations depended crucially, for example, on what sorts of clouds would grow under certain conditions. The fastest computers even today are far from able to calculate the details of every cloud on the planet. They must settle for figuring the average behavior of clouds in each cell of a grid, where the cell is hundreds of kilometers wide. Modelers had to develop "parameterizations," working up a set of numbers (parameters) to represent the net effects of all the clouds in a cell under given conditions. To get these numbers, the modelers had only some basic equations, fragments of unreliable data, and guesswork.

If that obstacle had somehow gone away, another remained. To diagnose the failings that kept models from matching the real world better, scientists needed an abundance of data—the actual profiles of wind, heat, moisture, and so on at many levels of the atmosphere and all around the globe. The data in hand through the 1960s were shamefully sparse. Smagorinsky put the matter succinctly in 1969: "We are now getting to the point where the dispersion of simulation results is comparable to the uncertainty of establishing the actual atmospheric structure."[10]

Help came from the drive to improve short-term weather predictions. The World Meteorological Organization (WMO) and the 1957–1958 International Geophysical Year (IGY) had made a good start, but they fell far short of gathering the kind of global data needed to truly understand the atmosphere. For example, even at the peak of the IGY there had been only one station reporting upper-level winds for a swath of the South Pacific Ocean that spanned one-seventh of the Earth's circumference. The lack of data posed insuperable problems for atmospheric scientists. Officials and scientists had managed to bring the problem to the attention of U.S. President John F. Kennedy, who saw an opportunity for enhancing his administration's prestige with a bold initiative. Addressing the United Nations General Assembly in 1963, Kennedy called for a cooperative international effort aimed at better weather prediction and, eventually, weather control. The WMO eagerly took up the proposal. The organization promptly launched the World Weather Watch, which coordinated thousands of professionals taking measurements in nearly every nation and on the high seas. The Watch has continued to the present day as the core WMO activity, unimpeded by the Cold War and other international conflicts.

By the 1970s computer models were predicting weather much better than the old rule-of-thumb forecasters, as far as three days ahead. That meant cash to farmers and other businesses, and the work attracted ample funding. The forecasting models required data on conditions at every level of the atmosphere at thousands of points around the world. Such observations were now being provided by the balloons and sounding rockets of the international World Weather Watch. Still better help came from outer space.

Use of satellites for weather "reconnaissance" had been proposed in a secret U.S. government report as early as 1950, and the first public satellite to monitor global weather had been built under a Department of Defense program and launched in 1960. Through the following decades, this program continued to build and operate secret meteorological satellites, using the exquisite and highly classified technologies developed for spy satellites. These technologies were gradually transferred to an open civilian program. When computer modelers reached the point where they could not progress without much better data on the actual atmosphere, the answer was *Nimbus 3*, launched in 1969. The satellite's infrared detectors could measure the temperature of the atmosphere comprehensively at various levels, night and day, over oceans, deserts, and tundra. Once again science was benefiting from money spent for practical military and civilian purposes. America's NASA (and later some other national space agencies) became the largest of all sources of money for climate research, if only because space technology was so expensive.

The computer work settled into steady, tenacious improvement of existing techniques. Modelers put in ever more factors, filling in the most gaping holes, and developed ever more efficient ways to use their rapidly improving computers. Encouragement came from a 1972 model by Mintz and Arakawa, which managed to roughly simulate the great changes as the sunlight shifted from season to season. During the next few years Manabe and his collaborators published a model that produced entirely plausible seasonal variations. That was a convincing test of the models' validity. It was almost as if a single model worked for two quite different planets— the planets Summer and Winter.

NASA's Goddard Institute for Space Studies in New York City took a different approach. The group working there had been developing a weather model as a practical application of its mission to study the atmospheres of planets. James (Jim) Hansen assembled a team to reshape their equations into a climate model for the Earth. By simplifying some features while adding depth to others, they managed to get a quite realistic-looking simulation that ran an order of magnitude faster than rival general circulation models. That permitted the group to experiment with multiple runs, varying one factor or another to see what changed. In such studies the global climate was beginning to look and feel to researchers like a comprehensible physical system, akin to the systems of glassware and chemicals that experimental scientists manipulated on their laboratory benches.

Sophisticated computer models were steadily displacing the traditional simplistic hand-waving models. In the digital models, it was clear from the outset that climate was the outcome of a staggeringly intricate complex of interactions and feedbacks among many global forces. Nobody had a straightforward explanation for even such a simple feature of the atmosphere as the semitropical doldrums. *In principle* the shape of the general circulation could be comprehended only indirectly, in the working through of a million calculations.

In their first decade or so of work, the modelers had a hard enough time just trying to understand a typical year's average weather. But in the mid-1960s a few of them had begun to take an interest in the question of what could cause changes in climate. They saw Keeling's curve of rising CO_2, and they saw Fritz Möller's discovery that simple models built out of a few equations showed disturbing instabilities. When Möller visited Manabe and explained his grotesque results, Manabe decided to look into how the climate system might actually change.

Manabe and his collaborators were already building a model that included, among many other things, the way air and moisture conveyed heat from the Earth's surface into the upper atmosphere. That was a big step beyond trying to calculate surface temperatures, as Möller and others since Arrhenius had attempted, just by considering the energy balance at the surface. To get a really sound answer, the atmosphere had to be studied as a tightly interacting system from top to bottom. In such a model, when the surface warmed, updrafts would carry heat by convection into the upper atmosphere—so the surface temperature would not run away as it had in Möller's model. The required computations were so extensive, however, that Manabe had to strip the model down to a single one-dimensional column, representing a slice of atmosphere averaged over the entire globe.

In 1967 Manabe's group used this model to test what would happen if the level of CO_2 in the atmosphere changed. Their target was something that would eventually become a central preoccupation of modelers: climate "sensitivity," that is, how much average global temperature would be altered by a given change in one variable (the Sun's output of energy, say, or the CO_2 level). They would run a model with one value of the variable (say, of CO_2), then run it again with a different value, and compare the answers. Ever since Arrhenius, researchers had pursued this question with highly simplified calculations. They used as a benchmark the difference produced by doubling the CO_2 level. After all, Keeling's curve showed that the level would probably double sometime in the twenty-first century. The number Manabe's group came up with for doubled CO_2 was a rise of global temperature of roughly 2°C (around 3° to 4°F). This was the first time a greenhouse-effect warming calculation included enough of the essential factors to seem reasonable to many experts. As Wally Broecker for one recalled, it was the 1967 paper "that convinced me that this was a thing to worry about."[11]

This model of a one-dimensional column of air was a far cry from a full, three-dimensional general circulation model. Using the column as a basic building block, Manabe and a collaborator, Richard Wetherald, constructed such a GCM in the early 1970s. It was still highly simplified. In place of the actual land and ocean geography, they pictured a planet that was half land and half swamp. But overall, this mock planet had a climate system that looked pretty much like the Earth's. In particular, the reflection of sunlight from the model planet at each latitude agreed pretty well with the actual numbers for the Earth, as measured by the new *Nimbus 3* weather satellite. For doubled CO_2 the computer predicted an average warming of around 3.5°C. When they published this in 1975,

Manabe and Wetherald warned that the result should not be taken too seriously; the model was still scarcely like a real planet.

One of the troubling complexities was explained by the prominent meteorologist William W. Kellogg. In 1975 he pointed out that industrial aerosols, as well as the soot from burning debris where forests were cleared, strongly absorbed sunlight—after all, smog and smoke are visibly dark. Hence, he argued, the main effect of human aerosols would be regional warming. Bryson and his co-workers continued to insist that smoke and haze had a powerful cooling effect—after all, they visibly dimmed sunlight. The debate was hard to resolve. Nobody could calculate from basic physics principles whether, under given circumstances, a haze would bring warming or cooling.

A still more vexing problem was cloudiness. If the planet got warmer, the amount of cloudiness would probably change, but change how? And what would a change in clouds mean for climate? Scientists were beginning to realize that clouds could either cool a region (by reflecting sunlight) or warm it (by trapping heat radiation from below). It depended on the type of clouds and how high they floated in the atmosphere. Worse still, it was becoming clear that the way clouds formed could be strongly affected by shifts in the haze of dust and chemical particles floating in the atmosphere. Little was known about how such aerosols helped or hindered the formation of different types of clouds.

This uncertainty was unacceptable, for people had begun to demand much more than a crude reproduction of the present climate. After the weather disasters and energy crisis of the early 1970s put greenhouse warming on the political agenda (for people who paid attention to technical issues), it became a matter of public debate whether the computer models were correct in their predictions of warming. Must we halt deforestation and turn away from fossil fuels? News reports featured disagreements among prominent scientists, especially over whether warming or cooling was likely. "Meteorologists still hold out global modeling as the best hope for achieving climate prediction," a senior scientist observed in 1977. "However, optimism has been replaced by a sober realization that the problem is enormously complex."[12]

Nevertheless, in the choice between warming and cooling, scientific opinion was coming down on one side. Perhaps in the natural course of events, over centuries the Earth would slide into an ice age—but the course of events was no longer natural. More and more scientists were coming to feel that greenhouse warming was the main thing to worry about. After all, rains washed most aerosols out of the lower atmosphere in a matter of weeks, but additional CO_2 would linger for several centuries. No matter whether aerosols warmed the Earth or cooled it, the greenhouse effect from increased CO_2 must dominate in the end. The sensational magazine tales of an ice age descending within the next century had reflected a bare handful of papers in the "peer-reviewed" scientific literature (that is, papers reviewed by other experts before publication). Several times as many scientific papers published in the 1970s were neutral, weighing the pros and cons of warming versus cooling. Still more came down on the side of future warming, and their numbers kept increasing.

Confidence could never come from a single line of attack, but several kinds of work pointed in the same direction. For example, Stephen Schneider and a collaborator studied the effects of dust by matching the temperature record of the past thousand years with volcanic eruptions. Their simplified model predicted that CO_2 warming would begin to dominate after 1980.

In 1977 the National Academy of Sciences weighed in with a major study by a panel of experts. They reported that temperatures might rise to nearly catastrophic levels during the next century or two. The panel's report, announced at a press conference during the hottest July the nation had experienced since the drought years of the 1930s, was widely noted in the press. Science journalists were growing better attuned to the views of climate scientists. When, in 1976, *Business Week* had explained both sides of the debate, it reported that "the dominant school maintains that the world is becoming cooler." Just one year later, the magazine declared that CO_2 "may be the world's biggest environmental problem, threatening to raise the world's temperature," with frightening long-term consequences.[13]

To get a more authoritative answer, the president's science adviser asked the National Academy of Sciences to study whether GCMs were trustworthy. The Academy appointed a panel, chaired by the veteran weather modeler Jule Charney and including other respected experts who had been distant from the recent climate debates. Their conclusion was unequivocal: the models were telling the truth. To make their conclusion more concrete, the panel decided to announce a specific range of numbers. Splitting the difference between Hansen's GCM, which predicted a 4°C rise for doubled CO_2, and Manabe's latest figure of around 2°C, the Charney panel declared they had rather high confidence that in the next century the Earth would warm up by about 3°C, plus or minus 50 percent, that is, 1.5°–4.5°C (2.7°–8°F). They concluded drily, "We have tried but have been unable to find any overlooked or underestimated physical effects" that could reduce the warming. "Gloomsday Predictions Have No Fault" was how *Science* magazine summarized the report.[14]

Many climate experts felt less confident. The computer models used toy planets scarcely resembling our Earth. They were flat geometrical constructions without mountains or other real geography, with wet surfaces in place of oceans, and clouds made up from guesses. There was no way to prove that some of the many pieces the models left out were not crucial. Journalists could track down experts who would give strong and definite statements about such questions, but most scientists were willing to live in uncertainty. They would watch as studies accumulated, seeing which ones reinforced or contradicted one another. On a subject as complex as climate change, no single finding would radically change anyone's opinion. One year a given expert might feel it was 60 percent likely that warming would come; after a few years that might shift down to 50 percent or up to 70 percent.

Faced with such doubt, the scientist's instinct is to ask what steps might be taken to bring more certainty. It is like adding pieces to a jigsaw puzzle with so many holes that you cannot yet make out what the overall picture looks like. Some of the holes would be hard to fill in. For instance, large-scale observation programs were needed, programs nobody was organizing or offering to pay for. As climate scientists grew increasingly worried about global warming, yet remained uncertain how likely it really was, they wanted more strongly than ever a centrally coordinated and generously funded program.

When a group of citizens (in this case scientists) decides that their government should do more to address some particular concern, they face a hard task. The citizens have only a limited amount of effort to spare, and officials are set in their bureaucratic ways. To accomplish anything—to bring about a new government program, for example—people must mount a concerted push. For a few years concerned citizens must hammer at the issue, informing the public and forging alliances with like-minded officials. These inside allies must form committees, draft reports, and shepherd legislation through the administration and legislature. Special interests that feel threatened by change will put up roadblocks, and the whole process is liable to fail from exhaustion. Typically, such an effort succeeds only when it can seize a special opportunity, usually news events that distress the public and therefore catch the eye of politicians.

In the early 1970s a few climate scientists had sought an opportunity to mount such a concerted push. They were spurred by the new calculations and data, which convinced them that climate might change sooner and more drastically than had seemed possible only a decade earlier. The media furor over crop failures and other environmental problems that erupted in the early 1970s gave climate scientists a chance to take action.

Most took a traditional route: they convened study groups and prepared reports for policy makers. The administration began to draft legislation to improve the organization and funding of the field. In 1976, with the recent droughts much in mind, a congressional committee began hearings. These were the first ever to address climate change as their main subject, starting what would become a long procession of scientists testifying that the rise of CO_2 was dangerous. Meanwhile, agency officials wrote and rewrote plans, negotiating tenaciously over who should get control of what research budget. The 1977 National Academy of Sciences report "Energy and Climate" kept up the pressure with its announcement that catastrophic warming might be in store.

The Academy's experts were by no means prepared to go further and recommend actual changes in the nation's energy policies, but they did drive home a general truth: the threat of climate change was intimately connected with energy production. As a front-page headline in the *New York Times* (15 July 1977) summed it up, "Scientists Fear Heavy Use of Coal May Bring Adverse Shift in Climate." Officials were starting to grasp the fact that CO_2 emissions had economic implications—and therefore political ones. The oil, coal, and electrical power industries began to pay attention.

Fossil-fuel policies were already under intense scrutiny. During the 1973 "energy crisis," inconvenience and anxiety beset millions of people as Persian Gulf states withheld their oil. When President Jimmy Carter's administration proposed to shift the United States from oil to coal, politics began to overlap scientific studies of climate change. The crisis gave a boost to advocates of nuclear energy, some of whom had been mentioning for years that if there ever was a greenhouse warming problem, nuclear reactors could solve it, since they burned no fossil fuels. The crisis was equally helpful to the nuclear industry's adversaries: advocates of renewable energy sources, ranging from federal solar-energy bureaucrats to anticorporate environmentalists. After all, more power generated by windmills would also mean less emission of CO_2. In the energy debates, however, climate change was only one more weight thrown into the balance. Compared with the many economic, political, and international issues, it was far from the heaviest weight in people's minds.

Nobody of consequence proposed to regulate CO_2 emissions or make any other significant policy changes to deal directly with greenhouse gases. Academy reports and other scientific pronouncements advised that any such action would be premature, given the lack of scientific consensus. The goals of the meteorological community and its friends in the bureaucracy remained the same: more money, and better organization, for research.

The spearhead of the effort was a Climate Research Board set up by the National Academy. The board's full-time chair, Robert M. White, was a widely admired scientist-administrator who had already served as head of the Weather Bureau and then of NOAA, as official representative to the World Meteorological Organization, as well as to various international meetings—one on whaling, for instance, another on desertification—and in countless other capacities. Bob White deserves notice as a fine example of many people whose names are not mentioned here, but whose contributions in administration and organization were indispensable.

At last in 1978 Congress passed the National Climate Act, which established the National Climate Program Office within NOAA. It was a step forward, more than other nations were doing, but the new office had a feeble mandate and a budget of only a few million dollars. Scientists did not get the well-coordinated research program they had called for. Without the backing of some unified community or organization, their movement had been impeded by the very fragmentation it sought to remedy.

In any case, lawmakers cared far more about the few years until the next election than about the following century. With the passage of the National Climate Act, the minor flurry of legislative attention ended. The program to study climate change was underfunded from the start. It won some large increases in the 1970s, but that ended in 1980. Such new money as was available seemed to go more into paperwork and meetings than into actual research.

Climate scientists in every nation had found it difficult to gain access to their respective policy makers. If they convinced their contacts among lower-level officials that there was a problem, these officials themselves had scant influence in higher reaches of government. The scientists discovered better opportunities when they turned to their peers in the international science community. Committees organized among a group of nations, pointing to the consensus of their prestigious scientists, might help convince each single nation's officials to respond. Besides, internationalization might offer some of the needed organization. After all, for something as global as weather, nobody could get far without exchanging information and ideas across national borders.

The natural place to turn would be the World Meteorological Organization, but the WMO was a federation of government weather agencies that brought together officials who were only loosely connected with academic scientists. The scientists had long since organized themselves in specialized societies such as the International Union of Geodesy and Geophysics, which sponsored conferences and other programs under the umbrella of the International Council of Scientific Unions (ICSU). Determined not to be left out in organizing weather research, ICSU negotiated with the WMO. In 1967 the two organizations jointly formed the Global Atmospheric Research Program (GARP). Its chief aim was to improve short-term weather prediction, but climate research was also on its agenda. Once the top experts on a GARP scientific committee had forged plans for a cooperative program, it was hard for the various national budget agencies to deny their scientists the funds needed to participate.

The chair of GARP's organizing committee during its crucial formative years was a Swedish climatologist, Bert Bolin. A savvy expert on subjects ranging from weather computation to the global carbon cycle, and even more admired for his skills as a team leader and diplomat, Bolin would be a mainstay of international climate organizing efforts for the next three decades.

Climate scientists met one another in an increasing number of international scientific meetings, from cozy workshops to swarming conferences. The 1971 meeting in Stockholm, "Study of Man's Impact on Climate," broke new ground with its stern warnings about the risk of future climate shocks (Chapter 4). The conference report became required reading for the delegates at the United Nations' first major conference on the environment, held the following year. Heeding the scientists' recommendations, the conference set in motion a vigorous program of cooperative research on the environment, including climate research. Meanwhile, the GARP committee set up a series of large-scale experiments that coordinated a great variety of government and academic institutions. An outstanding example was a project conducted in 1974, the GARP Atlantic Tropical Experiment (GATE, an acronym containing an acronym!). That summer, forty research ships and a dozen aircraft from twenty nations took measurements across a large stretch of the tropical Atlantic Ocean, studying the flows of heat and moisture into the atmosphere.

That was the easy part. However complex the oceans and atmosphere might be, research on them followed well-defined tracks. But the more scientists looked at the climate system from an international perspective, the more they noticed that there were additional components, even more complicated and scarcely studied at all. They began to discover evidence that the forests of Africa, the tundra of Siberia, and other living ecosystems were somehow essential parts of the climate system. How did they fit in?

Little was known about connections between the planet's biomass and the atmosphere. The few people who looked into the question found that the amount of carbon bound up in trees, peat bogs, soils, and other products of terrestrial life is several times greater than the amount in the atmosphere. The stocks of organic carbon seemed to have been fairly stable over millions of years. The likely cause of stability was a fact demonstrated by experiments in greenhouses and in the field: plants often grow more lushly in air that is "fertilized" with extra CO_2. Thus, if more of the gas was added to the atmosphere, it should be rapidly taken up, made into wood and soil. This was one more version of the argument that the atmosphere was automatically stabilized, part of the indestructible Balance of Nature.

Most geological experts thought that even on a lifeless planet, the atmospheric balance would remain stable. It seemed reasonable to expect that chemical cycles would long ago have settled down into some kind of equilibrium among air, rocks, and seawater. Compared with those titanic kilometer-thick masses of minerals, the planet's thin scum of bacteria and so forth hardly seemed worth considering. Thus, for example, a 1966 Academy of Sciences study of climate change concentrated on cities and industry; the panel remarked that changes to the countryside, such as irrigation and deforestation, were "quite small and localized," and it set that topic aside without study.[15]

As evidence mounted that global harm could be inflicted by such human products as chemical pesticides and dust, the traditional belief in the automatic stability of biological systems faltered. Concerns were redoubled by the African drought of the early 1970s.

Was the Sahara Desert expanding southward as part of a natural climate cycle that would soon reverse itself, or was something more ominous at work? For a century experts on Africa had worried that overgrazing could cause changes in the land: "man's stupidity" would create a "manmade desert."[16] In 1975 Charney proposed a mechanism: noting that satellite pictures showed widespread destruction of African vegetation from overgrazing, he pointed out that the barren clay reflected sunlight more than the grasses had. He figured this increase of albedo would make the surface cooler, and that might change the pattern of winds so as to bring less rain. Then more plants would die, and a self-sustaining feedback would push on to full desertification.

Charney was indulging in speculation, for computer models of the time were too crude to show what a regional change of albedo would actually do to the winds. It would be a few more years before models demonstrated that vegetation is indeed an important factor in a region's climate. But it didn't require detailed proof for scientists to grasp the truth of Charney's primary lesson. Human activity could change vegetation enough to affect the climate. The biosphere did not necessarily regulate the atmosphere smoothly, but could itself be a source of instability.

The public was meanwhile learning how slash-and-burn farming was eating up tropical forests, while the remnants of the great forests of North America were likewise shrinking away. Concern about these losses was rising—although for the sake of wildlife rather than climate. At the same time a few scientists pointed out that these forests were a significant player in global cycles of carbon and water. A forest evaporating moisture can be wetter than an ocean in its effect on the air above it. But just what kind of changes would deforestation bring? The answers lay in an uncharted no-man's-land between the fields of meteorology and biology.

Only a few things could be measured with confidence. Statistics compiled by governments on the use of fossil fuels told how much CO_2 was going into the atmosphere from industrial production. And Keeling, indefatigably pursuing his measurements decade after decade, showed how much of the gas remained in the air, nudging the curve higher year by year. The two numbers were unequal. Roughly half of the gas from burning fossil fuels was missing. Where was the missing carbon going?

There were only two likely places. The carbon must wind up either in the oceans or in biomass. In the early 1970s Broecker and others developed models for the movement of carbon in the oceans, including the carbon processed by living creatures. They calculated that the oceans were taking up much of the new CO_2, but not all of it. The residue must somehow be sinking into the biosphere. Perhaps trees and other plants were growing more lushly thanks to CO_2 fertilization?[17]

That was hard to check. Few solid studies of fertilization had been published by plant biologists—a type of specialist that had scarcely interacted with climate scientists. And as Keeling remarked, even with good data on past conditions, any calculation of the present or future fertilizer effect would be unreliable. Every gardener knows that giving a plant more fertilizer will promote growth only up to a certain level. Nobody knew where that level was if you gave more CO_2 to the world's various kinds of plants. "We are thus practically obliged to consider the rate of increase of biota as an unknown," Keeling warned.[18] Some rough calculations suggested that land plants might not be taking up carbon overall. The decay of organic matter in soils was increased by deforestation and other human works, so the land biota could be a major net *source* of greenhouse gas.

The uncertainties became painfully obvious at a workshop held in Dahlem, Germany, in 1976. Bolin argued that human damage to forests and soils was releasing a very large net amount of CO_2. Since the level in the atmosphere was not rising all that fast, it followed that the oceans must be taking up the gas much more effectively than geochemists like Broecker had calculated. George Woodwell, a botanist who studied ecosystems, went still further with his own calculations. He argued that deforestation and agriculture were putting into the air as much CO_2 as the total from burning fossil fuel, or maybe even twice as much. His message was that our attack on forests should be halted, not just for the sake of preserving natural ecosystems, but also to preserve the climate.

Broecker and his colleagues thought Woodwell was making ridiculous extrapolations from scanty data. Defending their own calculations, the oceanographers and geochemists insisted that the oceans could not possibly be taking up so much carbon. At the Dahlem conference the scientists debated one another vociferously. The arguments spilled over into social questions, all the issues of environmentalism and government intervention that deforestation and desertification posed. People's beliefs about the sources of CO_2 were becoming connected to their beliefs about what actions (if any) governments should take. Woodwell insisted that tropical deforestation and other assaults on the biosphere were "a major threat to the present world order."[19] He publicly called for a halt to burning forests, as well as aggressive reforestation to soak up excess carbon.

Such discussions were no longer restricted to scientists and the mid-level government officials they dealt with. Saving forests had become a popular idea in the growing environmental movement (in which Woodwell played a prominent role as an organizer). The forestry and fossil-fuel industries took note, recognizing that worries about greenhouse gases might lead to government regulation. They were joined by political conservatives, who tended to lump together all claims about impending ecological dooms as left-wing propaganda.

When environmental ideals had first stirred, around the time of Theodore Roosevelt, they had been scattered across the entire political spectrum. A traditional conservative, let us say a Republican birdwatcher, could be far more concerned about "conservation" than a Democratic steelworker. (In more recent times, Communist nations at the far end of the traditional Left became the planet's most ruinous polluters.) But during the 1960s, as a new Left rose to prominence, it became permanently associated with environmentalism. Perhaps that was inevitable. Many environmental problems, like smog, seemed impossible to solve without government intervention. Such interventions were anathema to the new Right that began to ascend in the 1970s.

By the mid-1970s conservative ideologues had joined forces with business interests to combat what they saw as mindless eco-radicalism. Establishing conservative think tanks and media outlets, they propagated sophisticated intellectual arguments and expert public-relations campaigns against government regulation for any purpose whatsoever. On greenhouse warming, it was naturally the fossil-fuel industries that took the lead. Backed up by some scientists, industry groups developed arguments ranging from elaborate studies to punchy advertisements, all aimed at persuading the public that there was nothing to worry about.

As environmental and industrial groups hurled uncompromising claims back and forth across a widening political chasm, most scientists found it hard to get a hearing for more ambiguous views. Journalists in search of a gripping story tended to present every scientific question as if it were a head-on battle between two equal and diametrically opposite sides. Yet most scientists saw themselves as just a bunch of people with various degrees of uncertainty, groping about in a fog.

Faced with the controversy over CO_2 emitted by deforestation, researchers tried to resolve the problem with data. In meetings and publications the experts wrangled, sometimes vehemently but always courteously. As occasionally happens in scientific debates, opinions tended to divide along disciplinary lines: oceanographers and geochemists versus biologists. The physical scientists like Broecker pointed out that their models of the oceans could be reliably calibrated with data on how the waters took up materials, especially the easily measured radioactive fallout from nuclear weapons tests. Woodwell's biology was manifestly trickier. His opponents argued that nobody really knew what was happening to the plants of the Amazon and Siberia. When he invoked field studies carried out in this or that patch of trees, his opponents brought up more ambiguous studies, or just said that studies of a few hectares here and there could scarcely be extrapolated to all the world's forests.

The key data finally came from measurements of radioactive carbon (using the fact that newly created isotopes cycled through the atmosphere and plants, whereas fossil-fuel emissions had long since lost their radioactivity). The ocean models turned out to be roughly correct. The gas emitted from decaying or burned plants was more or less balanced by the amount taken up by other plants. Perhaps deforestation was compensated for by more vigorous growth resulting from fertilization by the increased CO_2 in the atmosphere. Woodwell denied this, but other scientists gradually concluded that his claims were exaggerated. Eventually he had to concede that deforestation was not adding as much CO_2 to the atmosphere as he had thought. Much remained unexplained, however. Nobody was sure exactly how to balance the global carbon budget.

An important lesson remained. As a team headed by Broecker wrote in 1979, Woodwell's claims that destruction of plants released huge amounts of CO_2 had been a "shock to those of us engaged in global carbon budgeting." The intense reexamination triggered by the claim had called attention to "the potential of the biosphere."[20] From the late 1970s onward, it was clear that nobody could predict the future climate very well until scientists could say how the planet's living systems affected the level of CO_2. Of course, to answer that you would have to know how the biosphere itself would change if the climate changed. And to answer that you would have to know how the atmosphere would respond to changes in the oceans, and ice sheets, and more. The time when each specialty could make progress on its own was past. The international community of scientists would have to devise some social mechanism to coordinate their thinking.

At an International Workshop on Climate Issues, held under WMO and ICSU auspices in Vienna in 1978, the participants organized a World Climate Conference, which convened the following year in Geneva. Here the ideas of virtually all the world's important climate experts came up against one another. Well in advance, the conference organizers had commissioned a set of review papers inspecting the state of climate science in the various specialties, and these were circulated, discussed, and revised. Then more than three hundred experts from over fifty countries assembled to examine the review papers and recommend conclusions. Views about what might happen to the climate spanned a broad spectrum, yet the experts managed to reach a consensus. In their concluding statement, the scientists at the conference recognized a "clear possibility" that an increase of CO_2 "may result in significant and possibly major long-term changes of the global-scale climate." This cautious indication of an eventual "possibility" was hardly news, and it caught little public or political attention.

As the 1980s began, the possibility of greenhouse warming had become prominent enough to be included for the first time in public opinion polls. A 1981 survey found that over a third of American adults claimed they had heard or read about "the greenhouse effect." The news had spread beyond the small minority who regularly followed scientific issues. When pollsters explicitly asked people what they thought of "increased CO_2 in the atmosphere leading to changes in weather patterns," nearly two-thirds replied that the problem was "somewhat serious" or "very serious."[21]

Most of these citizens would never have brought up the subject by themselves. Only a small fraction understood that the risk of climate change was due mainly to carbon dioxide from fossil fuels. They blamed smog or other chemical pollution, nuclear tests, even spaceship launches. And those who worried most about the environment were seldom concerned with global affairs, directing their dismay instead at an oil spill or chemical wastes that endangered a particular neighborhood. Many people now suspected that they ought to be concerned about the greenhouse effect, but among the world's many problems it did not loom large.

THE ERRATIC BEAST

Ed Lorenz thought the climate might move in any direction, with little warning. His work in meteorology had helped lay the foundation of the newly fashionable chaos theory, and he continued to take the lead in studying how tiny initial variations could tilt a complex system this way or that. At a 1979 meeting he asked a famous question: "Does the flap of a butterfly's wings in Brazil set off a tornado in Texas?" His answer—perhaps it could—became part of the common understanding of educated people.[1]

Climate change had once seemed a simple enough concept, a gradual evolution responding to a direct push, whether your favorite theory said the push was a change in sunlight or volcanic haze or the level of CO_2. Decade after decade, scientists had discovered complications. The uncertain possibilities of aerosol pollution and deforestation were bad enough, but in the late 1970s and 1980s even more factors turned up. The climate began to look less like a simple mechanical system than like a confused beast that a dozen different forces were prodding in different directions.

Lorenz thought the outcome might be unpredictable in principle. He and others argued that the warming and cooling trends of the past century might not be evidence of responses to aerosols or a greenhouse effect or anything else in particular. It might be only the beast erratically lunging this way and that, in obedience to its incalculably complex internal reaction to the various external pressures.

Most scientists agreed that climate has features of a chaotic system, but they did not think it was wholly random. It might well be impossible in principle to predict that a tornado would hit a particular town in Texas on a particular day (not because of one guilty butterfly, of course, but as the net result of countless tiny initial influences). Yet tornado seasons came on schedule. That type of consistency showed up in the computer simulations constructed in the 1980s. Start a variety of GCM runs with different initial conditions, and they would show random variations in the weather patterns in different regions and different years. But the runs would converge when it came to the global temperature, averaged over a few years for all regions. And every model ended with some sort of warming during the next century.

This was not good enough for the critics, who noted many points where all the models shared uncertain assumptions and flimsy data. The modelers admitted that they still had a long way to go. Their arcane GCMs could inspire little confidence among the public. People wanted a more straightforward indicator—like the weather outside their windows. It was scarcely possible to get the public, or even most scientists, to take greenhouse warming seriously when the average temperature of the planet was dropping.

But was it? In 1975 two New Zealand scientists reported that while the Northern Hemisphere had been cooling over the past thirty years, their own region, and probably other parts of the Southern Hemisphere, had been warming. There were too few weather stations in the vast, unvisited southern oceans to be certain, but other studies tended to confirm it. The cooling since around 1940 had been observed mainly in northern latitudes. Perhaps the greenhouse warming was counteracted there by cooling from industrial haze? After all, the Northern Hemisphere was home to most of the world's industry. It was also home to most of the world's population, and, as usual, people had been most impressed by the weather where they lived.

Scientists and governments needed to know for sure what had been happening to the weather. Thousands of stations around the world had churned out daily numbers for up to a century, but the numbers conformed to no single standard: they made an almost indecipherable muddle. Around 1980 two groups undertook to work through the numbers in all their messy historical and technical details, rejecting unreliable sets of data and tidying up the rest.

First to weigh in was Jim Hansen's group in New York. It reported that "the common misconception that the world is cooling is based on Northern Hemisphere experience to 1970." Just around the time that meteorologists had noticed the cooling trend, it had apparently reversed. From a low point in the mid-1960s, by 1980 the world as a whole had warmed some 0.2°C.[2] That was what would be expected as the steady buildup of greenhouse gas overcame the temporary effects of aerosols, especially as some nations imposed pollution controls. (Moreover, there had been more volcanic emissions than usual during the period.) The temporary northern cooling from the 1940s through the 1960s had been bad luck for climate science. By feeding skepticism about the greenhouse effect, and by provoking some scientists and many journalists to speculate publicly about the coming of a new ice age, the cool spell gave the field a reputation for fecklessness that it would not soon live down (Figure 2).

Any greenhouse warming would be masked not only by random natural variations and industrial pollution, but also by some fundamental planetary physics. If anything added heat to the atmosphere, much of the heat energy would soon work into the upper layer of the oceans. The expected rise of temperature as measured in the air would be delayed for a few decades while the oceans warmed

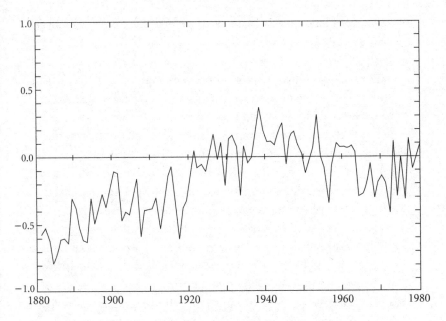

Figure 2. THE ERRATIC RISE OF TEMPERATURE.

The average surface temperature of the Northern Hemisphere (shown as differences from the 1946–1960 mean in °C). This 1982 graph by a British group shows the prominent rise to the 1940s, confusing decline through the 1960s, and vague beginnings of the subsequent rise (which soared much higher in the 1990s; see Figure 3, p. 182). (P. D. Jones et al., *Monthly Weather Review* 110 [1982]: 67, reproduced by permission of American Meteorological Society.)

up. The Charney Committee of experts had explained the effect in 1979: "We may not be given a warning until the CO_2 loading is such that an appreciable climate change is inevitable."[3] When would the warning come? In 1981 Hansen's group boldly predicted that, considering how fast the gas was accumulating, "carbon dioxide warming should emerge from the noise level of natural climatic variability" by the end of the century. Other scientists, using different methods of calculation, got similar results. The discovery of global warming—that is, plain evidence that the greenhouse effect really operated as predicted—would come sometime around the year 2000.[4]

The second group analyzing global temperatures was the British government's Climatic Research Unit at the University of East Anglia, led by Tom M. L. Wigley and P. D. Jones. In 1986 they produced the first entirely solid and comprehensive global analysis of average surface temperatures. The warmest three years in their 134-year record had all occurred in the 1980s. Hansen and a collaborator made an analysis using different methods and came up with a parallel result. It was true: an unprecedented warming was under way.

In such publications, the few pages of text and graphs were the visible tip of a prodigious, unseen volume of work. Many thousands of people in many countries had spent much of their lives measuring the weather, while thousands more had devoted themselves to organizing and administering the programs, improving the instruments, standardizing the data, and maintaining the records in archives. In geophysics not much came easily. One simple sentence (like "last year was the warmest year on record") might be the distillation of the labors of a multigenerational global community. And it still had to be interpreted.

Most experts saw no solid proof that serious warming lay in the future. After all, reliable records covered little more than a century, and they showed large fluctuations. Couldn't the current trend be just another temporary wobble? Schneider, one of the scientists least shy about warning of climate dangers, acknowledged that "a greenhouse signal cannot yet be said to be unambiguously detected in the record." Like Hansen and others, he predicted that the signal would not emerge clearly until around the end of the century.[5]

The way temperatures had soared in some decades but dipped in others showed that factors besides the greenhouse effect must be at work. When modelers like Hansen added a factor for volcanic eruptions to the rise of CO_2, they found this could explain a good part of the variation—but not all. What else was pushing on climate?

There were hints that part of the variation followed a regular cycle. For example, Dansgaard's cores of ancient ice drilled from deep in the Greenland ice sheet showed a cycle about eighty years long. The Danish team supposed the Sun was responsible, for the number of sunspots seemed to follow a similarly long cycle. Broecker now suggested that the low temperatures of the 1960s were caused by a downtrend in this solar cycle, even more than volcanoes and industrial haze. In 1975 he published an influential article suggesting that once the solar cycle turned, the world might get a "climatic surprise," a serious rise of temperature that he called, in a phrase new to most people, "global warming."[6]

Later studies found that Dansgaard's cycles, insofar as they existed at all, seemed to represent something that happened in the North Atlantic Ocean, not on the Sun. It was just another case of supposed global cycles that faded away as more data came in. It was also one of several cases in which Broecker's scientific instincts were better than his evidence, for other reasons appeared to point a finger at solar variations as an influence on climate.

Back in the early 1960s one of the new radiocarbon experts, Minze Stuiver, had collaborated with Hans Suess to show that the amount of radiocarbon in ancient tree rings varied from century to century. Many scientists were upset, for the erratic variations could undermine the technique of radiocarbon dating. But what looks like unwelcome noise to one specialist may contain information for another. Stuiver noted that radiocarbon is generated in the atmosphere by cosmic rays from the far reaches of the universe, and he pointed out that the magnetic field of the Sun impedes the flux of cosmic rays reaching the Earth. Perhaps the radiocarbon record said something about changes on the Sun?

In 1965 Suess tried correlating the new data with weather records, in the hope that radiocarbon variations "may supply conclusive evidence regarding the causes for the great ice ages."[7] He focused on a bitter cold spell reported in European weather records of the sixteenth and seventeenth centuries: the Little Ice Age, when crops had failed repeatedly and the Thames at London froze solid in winter. That had been a period of relatively high radiocarbon. Casting a sharp eye on historical sunspot data, Suess noticed that around the time of the Little Ice Age, few sunspots had been recorded. Fewer sunspots indicated a lower level of solar magnetism, which meant more cosmic rays could get through, which created more radiocarbon. In short, Suess suggested, more radiocarbon pointed to changes on the Sun that somehow connected with cold winters.

Some found the connections plausible, but to most scientists the speculation sounded like just one more of the countless sunspot correlations that had been announced, only to fail sooner or later. Even if the evidence had been stronger, it would have met with deep skepticism, for scientists cannot well fit data into their thinking unless theory meanwhile prepares a place. The variations of sunspots and cosmic rays were negligible compared with the Sun's total output of energy. How could such trivial variations possibly have a noticeable effect on climate?

In 1975 the respected meteorologist Robert Dickinson took on the task of reviewing the American Meteorological Society's official statement about solar influences on weather. He concluded that such influences were unlikely, for there was no reasonable mechanism in sight—except, maybe, one. Perhaps the electrical charges that cosmic rays brought into the atmosphere somehow affected the way cloud droplets condensed on dust particles? Dickinson hastened to point out that this was pure speculation. Scientists knew very little about how clouds formed; they would need to do much more research "to be able to verify or (as seems more likely) to disprove these ideas."[8] For all his skepticism, Dickinson had left the door open a crack.

In 1976 a solar physicist tied all the threads together in a paper that soon became famous. John (Jack) Eddy was one of several solar physicists in Boulder. Despite the variety of climate experts in that city, Eddy was ignorant of the radiocarbon research—an example of the poor communication between fields that had always impeded climate studies. Eddy was having scant success in the usual sort of solar physics research, and in 1973 he lost his job as a researcher, finding only temporary work writing a history of NASA's *Skylab*. In his spare time he pored over much older history. He was reviewing centuries of naked-eye sunspot records, aiming for a definitive confirmation of the long-standing belief that the Sun is comfortably stable.

He found, however, that the sunspot cycle did not look constant at all. This discovery, like many in science, was not entirely new. An almost complete absence of sunspot observations during the Little Ice Age had been noticed not only by Suess but by several others before him. In particular, back in 1890 a British astronomer, E. Walter Maunder, had drawn attention to the evidence and a possible climate connection. Other scientists had thought this was just another case of dubious numbers at the edge of detectability, and Maunder's publications sank into obscurity. A scientific finding cannot flourish in isolation; it needs support from other findings.

"As a solar astronomer I felt certain that it could never have happened," Eddy later recalled.[9] But hard historical work gradually persuaded him that the early modern solar observers were reliable— the absence of sunspot evidence really was evidence of an absence. Other scientists were skeptical, but as Eddy pushed his arguments, he learned of the radiocarbon evidence. Stuiver and others had meanwhile confirmed the connection between solar activity and the radiocarbon in tree rings and other fossil sources. When Eddy presented his full results in 1976, correlating sunspots, radiocarbon, and temperatures, many found the evidence persuasive.

However strong a scientific finding may seem, if it is surprising, scientists will demand confirmation. Schneider, Hansen, and others found that they could indeed get a better match to past temperature trends if they took into account not only the sporadic cooling due to dust from volcanic eruptions, but also the solar variability indicated by sunspots and radiocarbon. Adjusting the strength of the presumed solar influence to match the historical temperature curve was guesswork, dangerously close to fudging. But sometimes a scientist must "march with both feet in the air," assuming a couple of things at once in order to see whether it all eventually works out. The results looked good enough to encourage further studies. Some of these failed to find any correlations. As a reviewer commented in 1985, "This is a controversial topic," and the connection between solar variations and climate change remained "an intriguing but unproven possibility."[10]

If changes on the Sun itself had been overlooked, what other pieces of the climate puzzle might be missing? Another, quite different set of scientific specialties added an entirely new type of question. The history of climate science is full of unexpected linkages, but perhaps none so odd and tenuous as the events that drew attention to trace chemicals. The inquiry began with the concern over how pollution from supersonic transport airplanes might alter the stratosphere (Chapter 4). In 1973 Mario Molina and Sherwood Rowland thought it would be interesting to take a look at what other chemical emissions might do. They were astonished to find that the minor industrial gases known as CFCs (chlorofluorocarbons) could have serious effects.

Experts had thought that CFCs were environmentally sound. They were produced in relatively small quantities. And they were extremely stable, never reacting with animals and plants. It turned out that the stability itself made CFCs a hazard. They would linger in the air for centuries, and eventually, Molina and Rowland realized, some would drift up into the stratosphere. There ultraviolet rays activated them and they became catalysts in a process that destroyed ozone. The high, thin layer of ozone blocks the Sun's ultraviolet rays, so removing this layer would result in an increase of skin cancers and probably still worse dangers to people, plants, and animals.

CFCs were the propellants in aerosol sprays: every day millions of people were adding to the global harm as they used cans of deodorant or paint. Science journalists alerted the public, and environmentalists jumped on the issue. Industries fought back with public relations campaigns that indignantly denied there was any risk whatsoever. Unconvinced, citizens bombarded government representatives with letters, and the U.S. Congress responded in 1977 by banning the chemicals from aerosol spray cans. The issue had no visible connection with climate. But it sent a stinging message about how fragile the atmosphere was, how easily harmed by human pollution. And it showed that technical scientific findings about a future atmospheric risk could arouse the public enough to sway legislation and strike at major industries.

Rowland and Molina's work on CFCs meanwhile provoked Veerabhadran ("Ram") Ramanathan of NASA to take a closer look at these unusual molecules. In 1975 he reported that CFCs absorb infrared radiation prodigiously—they were greenhouse gases. A simple calculation suggested that CFCs, at the concentrations they would reach by the year 2000, all by themselves might raise global temperature by 1°C. Other scientists followed up with a calculation on other gases that had previously been little considered: methane (CH_4, a main component of "natural gas") and nitrates (especially the N_2O that is emitted when fertilizer is spread about). If the level of these gases in the atmosphere doubled, that could raise the temperature another 1°C. In 1985 Ramanathan led a team that looked at some thirty trace infrared-absorbing gases. The additional greenhouse gases, the team estimated, together could bring as much global warming as CO_2 itself. For climate scientists that was a bombshell. Global warming might be coming toward them twice as fast as they had expected!

The trace gases had been overlooked because the amounts of them in the atmosphere were minuscule compared with CO_2. But there was already so much CO_2 in the air that the spectral bands where it absorbed radiation were mostly opaque already. You had to add a lot more CO_2 to make a serious difference. A few moments' thought would have told any physicist that it was otherwise for trace gases. For those, each additional wisp would help obscure a "window," a region of the spectrum that until then had let radiation through unhindered. But the simple is not always obvious unless someone points it out. Understanding took a while to spread. Government officials and even most scientists persisted in thinking that "global warming" was essentially synonymous with "increasing CO_2." Meanwhile, many thousands of tons of other greenhouse gases were pouring into the atmosphere.

Some scientists did now recognize the importance of methane and looked into its role in global carbon cycles. Methane is the swamp gas that bubbles out of bogs; it comes mostly from the bacteria that thrive everywhere from garden soil to the guts of termites. These natural emissions were much greater than the amount of methane that escaped as humans extracted and burned natural gas. But that did not mean human effects were negligible. Humanity was imposing its will on most of the world's fertile surface, transforming the entire global biosphere. Specialists in obscure fields of research turned up a variety of methane sources that were rapidly increasing. The gas was emitted in geophysically significant quantities by the bacteria found, for example, in the mud of rice paddies and in the stomachs of the proliferating herds of cows. And what about accelerated emissions from the soil bacteria that flourished under deforestation and the advance of agriculture?

The effects turned out to be huge. In 1981 a group reported that methane in the atmosphere was increasing with astounding speed. A study of air trapped in bubbles in cores drilled from the Greenland ice cap confirmed that a rise of methane had begun a few centuries ago, and it had wildly accelerated in recent decades. Through painstaking collection of air samples, the recent rise was measured accurately by 1988. The methane level had increased 11 percent in the previous decade alone. And each molecule of the gas had a greenhouse effect about twenty times that of a molecule of CO_2.

That raised alarming possibilities for feedbacks. A huge reservoir of carbon is frozen in the deep permafrost layers of peat that underlie northern tundras. As Arctic regions warmed up, would the endless expanses of sodden tundra emit enough methane to accelerate global warming? Even more ominous were the enormous quantities of carbon locked in the strange "clathrates" (methane hydrates), icelike substances found in the muck of seabeds around the world. The clathrates are kept solid only by the pressure and cold of the overlying water. In the early 1980s a few scientists pointed out that if a slight warming penetrated the sediments, the clathrates might melt and release colossal bursts of methane and CO_2 into the atmosphere—which would bring still more warming.

Of course, there would also be feedbacks from normal biological production, as warming altered the amounts of gases emitted or taken up by forests, grasslands, ocean plankton, and so forth. The only way to predict the future of such a complex system would be through computer modeling. A great deal had to be learned, however, before anyone could write equations that would accurately represent all the important effects.

To observe the climate working and changing as a single complex system, there was nothing so useful as the problem that had opened up the question in the first place: the ice ages. Understanding those mighty swings would take us a long way toward understanding the climate system. The prospects were good, for a crisp synopsis of past changes was emerging from continually improved studies of seabed clay and glacial ice. Technologies for working in the open ocean had been developed extensively for commercial purposes such as oil prospecting. In the 1970s these technologies were put to use in the Deep Sea Drilling Project. This series of cruises, funded by the U.S. National Science Foundation, pulled up long cores from ocean floors around the world. The sequence of temperatures recorded in the layers of clay was found to go up and down in fine agreement with the records from the ice of Greenland and Antarctica. Researchers began to combine all these data in a single discussion.

The finest core of all was extracted by a French-Soviet team at the Soviet Vostok Station in Antarctica. It was a truly heroic feat of technology, wrestling with drills stuck a kilometer down at temperatures so low that a puff of breath fell to the ground in glittering crystals. Vostok was the most remote spot on the planet, supplied once a year by a train of vehicles that clawed across hundreds of kilometers of ice. Underfunded and threadbare, the station was fueled by the typically Russian combination of cigarettes, vodka, and stubborn persistence. ("What do you do for recreation?" "Wash . . . you have a bath once every ten days.")[11] In the late 1980s, when they reached bottom and hauled up their core, they found it stretched back over 400,000 years—through four complete cycles of glaciation.

For two decades, group after group had cut samples from cylinders of ice in hopes of measuring the level of CO_2 trapped within the tiny air bubbles. Every attempt had failed to give plausible results. Finally, in 1980, reliable methods were developed. The trick was to clean an ice sample scrupulously, crush it in a vacuum, and quickly measure what came out. The Vostok results were definite, unexpected, and momentous.

In each glacial period, the level of CO_2 in the atmosphere had been lower than during the warm periods in between—lower by as much as 50 percent. Nobody could explain what caused the level to soar and plunge so greatly as glacial periods came and went. The Vostok core tipped the balance in the greenhouse-effect controversy, nailing down an emerging scientific consensus: the gas did indeed play a central role in climate change. This work fulfilled the old dream that studying the different climates of the past could be almost like putting the Earth on a laboratory table, switching conditions back and forth and observing the consequences.

The ice-core results answered the old objection to Milankovitch's orbital theory of ice ages: if their timing was set by variations in the sunlight falling on a given hemisphere, why didn't the Southern Hemisphere get warmer as the Northern Hemisphere cooled, and vice versa? The answer was that changes in atmospheric CO_2 (and methane, which likewise soared and plunged) physically linked the two hemispheres, warming or cooling the planet as a whole. The findings also suggested how the feeble shifts in sunlight of the 100,000-year cycle could raise and destroy continental ice sheets. It seemed there were indeed positive feedbacks: a little warming caused something to emit greenhouse gases into the atmosphere, and that made for more warming, and so on. The little orbital shifts only tipped the balance to set a powerful process in motion.

What was the feedback mechanism? As the oceans got warmer, some of their CO_2 would evaporate into the air, but that was not enough. The emission of gases from warmed-up wetlands and forests was another likely candidate, and there were more. Scientists suggested various possibilities, each more peculiar than the last. For example: in glacial ages the planet had bigger deserts and stronger winds; these put more dust into the air (the dust layers were visible in ice cores); the minerals in such dust could provide essential fertilization for ocean plankton, which perhaps would bloom prodigiously; the creatures would take up CO_2 from the air; when they died and drifted to the seafloor, the carbon would be buried with them, reducing the greenhouse effect; the glacial age would deepen. Or perhaps something else had happened—there was no lack of ingenuity in finding other possible feedback mechanisms.

Dust remained a central puzzle for the present climate, too. The arguments over whether aerosol particles warmed or cooled the planet were intricate and perplexing. For example, historical research on volcanoes had turned up a distinct pattern of global cooling for a few years following each major eruption. But the visible smoke of an eruption was mainly tiny particles of volcanic glass, which dropped out of the air within a few weeks. How could that be responsible for such long-term effects? The atmosphere also rapidly rid itself of the mineral dust from soils degraded by human agriculture. Likewise, it disposed of the carbon soot in smoke from factories and from slash-and-burn forest clearing. How could these make more than a temporary and local difference?

The answer was hidden in something else thrown into the air. Anyone looking at city smog—or smelling it—might guess that simple chemical molecules were a main component. The studies of smog that began in the 1950s brought a few scientists to study molecules in the air, and they found that one of the most important was sulfur dioxide, SO_2. Emitted profusely by volcanoes as well as by industries burning fossil fuels, SO_2 rises into the atmosphere and combines with water vapor to form minuscule droplets and crystals of sulfuric acid and other sulfates. In the early 1970s, when worry about the ozone layer had driven scientists to study the chemistry of the stratosphere, they found that the most significant aerosols there were sulfates. These lingered for years, reflecting and absorbing radiation. Did that make any difference? A hint came from the clouds covering the planet Venus. In the early 1970s precise telescope observations identified the source of the mysterious haze that helped make the planet a greenhouse hell. The haze was mainly sulfates.

Haze on the Earth, outside the smoggy cities, was commonly assumed to be a "natural background" from soil particles, sea salt crystals, volcanic smoke, and so forth. That was challenged in 1976 by Bert Bolin and Robert Charlson. Analyzing air purity data collected by government agencies, they showed that sulfate aerosols measurably dimmed the sunlight across much of the eastern United States and Western Europe. Among all the aerosols arising from human activity, they calculated, sulfates played the biggest role for climate. The effect at that time seemed minor. But with consumption of fossil fuels rising, anyone who wanted to calculate the long-term future of climate would somehow have to take sulfates into account.

A few computer-modeling groups took up the challenge. Especially convincing was a paper Hansen's team published in 1978. They looked back to 1963, when the eruption of Mt. Agung in Indonesia had thrown some three million tons of sulfur into the stratosphere. A simplified model calculated that the sulfates should have made for cooling, and the numbers matched in all essential respects the global temperature changes actually observed in the mid-1960s. Contrary to what some scientists had argued, and unlike what happened on Venus, it seemed that the net effect of sulfates on Earth was to cool the surface.

Yet this was far from proving that the net effect of all human pollution would make for cooling. The arguments so far had addressed only how aerosols directly intercepted radiation. But since the 1960s a few scientists had pointed out that this might not be their most important effect. New observations showed that under natural conditions, there were often too few nuclei to help water droplets coalesce into clouds. This was the kind of thing Walter Orr Roberts had talked about when he had pointed to cirrus clouds evolving from jet contrails. That had been taken as a temporary, local effect. Now some wondered whether human emissions, by adding nuclei for water droplets, might increase cloudiness worldwide. If so, what would that mean for climate?

In 1977 Sean Twomey cast some light into these shadows. He showed that reflection of sunlight from clouds depends in a complex way on the number of nuclei. Depending on temperature, humidity, and the type and number of particles, aerosols might bring a thin mist, or a thick cloud, or rain followed by clear skies. So adding aerosols could either raise or lower cloud reflectivity, depending on a variety of factors. Moreover, while a cloud would reflect sunlight back into space, it would also intercept radiation coming up from below, causing a greenhouse effect. Extensive calculation was needed to figure whether the net effect of a particular sort of cloud was cooling or warming. Twomey calculated that overall, the effect of human aerosols should be to cool the Earth.

Other scientists paid little attention. The theory was too tricky for anyone to trust Twomey's calculations very far. And there were no convincing observations to pin down the equations—even the effects of seeding clouds with silver iodide smoke remained controversial despite decades of costly experiments. Hardly any measurements had been made of the actual mixture of particles and chemical molecules that drifted in the atmosphere. Still less was known about the ways that the chemicals interacted with one another, although studies of smog showed that this was crucial. Few researchers cared to stake years of their professional lives on such hideously complex, perhaps insoluble problems.

Yet the problems became steadily more unavoidable. Even in the Arctic, where the immense empty landscapes promised only pristine air, scientists were startled to find a visible haze of pollutants drifting up from industrial regions. They were coming to recognize that humans were now the dominant source of the atmosphere's sulfate aerosols. In 1987 a dramatic visible demonstration convinced many scientists that Twomey's calculations deserved attention. Satellite pictures of the oceans displayed persistent clouds above shipping lanes, a manifest response to smoke from the ships. Apparently aerosols did create enough clouds to reflect sunlight significantly. So what did that mean for the future climate? The few people dedicated to the troublesome research were far from providing an answer.

Research moved ahead faster on an equally difficult but more obviously central feature of the climate system. "We may find that the ocean plays a more important role than the atmosphere in climatic change," a panel of experts remarked in 1975.[12] The first generation of general circulation models had treated the oceans as if they were simply a wet surface. But ocean currents carried a huge amount of heat from the tropics to the poles. That was a key component of the engine of climate, a component that the GCMs—models of the general circulation of the atmosphere—had not tried to include.

Two obstacles kept modelers from handling the oceans in the same way as the atmosphere. First, although the atmosphere was measured daily in thousands of places, oceanographers had only occasional and scattered data. The sporadic expeditions, retrieving bottles of water here and there from kilometers down, were like a few blind men crawling about a vast prairie. Second, models of the atmosphere could bypass many difficulties by letting a simple equation or average number stand in for the complexities of a swirling storm; but analogous processes in the seas, the heavy, decades-long sloshing of water about an ocean basin, had to be computed in full detail. The fastest computers of the 1970s lacked the capacity to calculate central features of the ocean system. They could not handle even something as fundamental and apparently simple as the vertical transport of heat from one layer to the next.

Oceanographers were coming to realize that crucial energy transfers were carried by myriad whorls of various sizes. At one extreme were microscopic circuits that shuttled heat down from the surface in some fashion nobody had yet figured out. At the other extreme were eddies bigger than Belgium that plowed through the seas for months. These colossal slow whorls were discovered only in the 1970s, thanks to an international study of the North Atlantic carried out by six ships and two aircraft. To the amazement of oceanographers, most of the energy in the ocean system was carried about by these eddies, not by ocean-spanning currents like the Gulf Stream. Calculating all the big and little whorls, like calculating individual clouds, was far beyond the reach of the fastest computer. Modelers had to work out parameters to summarize the main effects, only this time for objects that were much harder to observe and less well understood than clouds. Even with gross simplifications, to get anything halfway realistic took many more numerical computations than the atmosphere required.

Suki Manabe had shouldered the task in collaboration with Kirk Bryan, an oceanographer with training in meteorology who had been recruited to the group to build a stand-alone numerical model of an ocean. The two teamed up to construct a model that coupled this ocean model to Manabe's atmospheric GCM. Manabe's winds and rain would help drive Bryan's ocean currents, while in return Bryan's sea-surface temperatures and evaporation would help determine the circulation of Manabe's atmosphere. In 1968 they completed a heroic computer run, some 1,100 hours long (totaling over twelve full days devoted to the atmosphere and thirty-three to the ocean).

Bryan wrote modestly that "in one sense the . . . experiment is a failure."[13] Even after running for a simulated century, the deep-ocean circulation had not nearly reached equilibrium. It was not clear what the final climate solution would look like. Yet it was a great success just to get a linked ocean-atmosphere computation that was at least starting to settle into an equilibrium. The result looked like a real planet—not our Earth, for in place of geography they used a radically simplified geometrical sketch, but with plausible ocean currents, trade winds, deserts, snow cover, and so forth.

Our actual Earth was poorly observed, but in the simulation one could see exactly how air, water, and energy moved about.

Following up, in 1975 Manabe and Bryan published results from the first coupled ocean-atmosphere GCM that had a roughly Earth-like geography. The supercomputer ran for fifty straight days, simulating movements of air and sea over nearly three centuries. At the end their simulated world-ocean still failed to show a full circulation. But the results were getting close enough to reality to encourage them to push ahead. Meanwhile, Warren Washington's team in Boulder developed another ocean model, based on Bryan's, and coupled it to their own quite different GCM. Their results resembled Manabe and Bryan's, a gratifying confirmation.

Ocean modeling was becoming a recognized specialty. A research program that had once seemed "a lonely frontier like a camp of the Lewis and Clark Expedition," as an ocean modeler recalled in 1975, took on "more of the character of a Colorado gold camp."[14] One reason was the breathtaking advances in computers. Equally important was a splendid addition to the limited stock of oceanographic data. In a major project that the U.S. government funded in the 1970s, the Geochemical Ocean Sections Study (GEOSECS), teams of researchers sampled seawater at many points. Their chief interest was the radioactive carbon, tritium, and other debris spewed into the atmosphere by nuclear bomb tests in the late 1950s. The fallout had landed on the ocean surface around the world and was gradually being carried into the depths. Thanks to its radioactivity, even the most minute traces could be detected. The bomb fallout "tracers" gave enough information to map accurately, for the first time, all the main features of the three-dimensional ocean circulation. At last modelers had a realistic target to aim at.

Various teams improved their ocean-atmosphere models through the 1980s, occasionally checking how they reacted to increased levels of CO_2. The results, for all their limitations, said something about the predictions of the earlier atmosphere-only GCMs. It turned out, as expected, that the oceans would delay the appearance of global warming for a few decades by soaking up heat. As Hansen's group warned, a policy of "wait and see" might be wrongheaded, for a temperature rise might not become apparent until much more greenhouse-effect warming was inevitable.[15] Aside from that, linking a somewhat realistic ocean to GCMs did not turn up anything to alter the predictions in hand for future warming.

There was a problem, however, that few appreciated. The computer models showed a steady, gradual change of climate as CO_2 increased. But modelers had adjusted the very structure of their jittery models until they changed smoothly rather than veering off into some impossible state. In the real world, when you push on something steadily it may remain in place for a while, then move with a jerk. Since the 1960s scientists had suspected that the climate system could shift in that abrupt way. In the 1980s disturbing new evidence confirmed it.

Hints of rapid climate change had shown up in the long core drilled in the 1960s at Camp Century in Greenland, but a single record could be subject to all kinds of accidental errors. Dansgaard's group built a new drill and went to a second location, some 1,400 kilometers distant from Camp Century, where they extracted gleaming cylinders of ice 10 centimeters in diameter and in total over 2 kilometers long. They cut out 67,000 samples, and in each sample analyzed the ratios of oxygen isotopes. The temperature record showed jumps that corresponded closely to the jumps at Camp Century.

In 1984 Dansgaard reported that the most prominent of the "violent" changes corresponded to the Younger Dryas oscillation, "a dramatic cooling of rather short duration, perhaps only a few hundred years."[16] Corroboration came from a group working under Hans Oeschger. An ice-drilling pioneer, Oeschger was now analyzing layers of lake-bed clay near his home in Bern, Switzerland. That was far indeed from Greenland, but his group found "drastic climatic changes" that neatly matched the ice record.[17]

Many felt that such large changes had to be regional, perhaps affecting the North Atlantic and Europe but not the whole planet. A look at North American and Antarctic records did not find the same features. Yet as ice drillers improved their techniques, they found, to everyone's surprise, large steps not only in temperature but also in the CO_2 concentration. Since the gas circulates through the atmosphere in a matter of months, the steps seemed to reflect abrupt worldwide changes. Oeschger was particularly struck by a rapid rise of CO_2 that others reported finding in Greenland ice cores at the end of the last ice age.

The main reservoir of carbon was the oceans, so that was the first place to think about. In 1982 Broecker visited Oeschger's group in Bern and explained current ideas about the North Atlantic circulation. Oeschger pondered how the carbon balance of the oceans might be changed, but he could not come up with a convincing mechanism. In fact, scientists later realized that the rapid variations seen in the ice cores were merely an artifact. They did not reflect changes in atmospheric CO_2, but only changes in the ice's acidity caused by dust layers; something had indeed changed swiftly, but not necessarily the CO_2 level. Yet the error had served a good purpose, for Oeschger's speculations had set Broecker to thinking.

Ever since the days when he had trudged around fossil lake basins in Nevada for his doctoral thesis, Broecker had been interested in sudden climate shifts. The reported jumps of CO_2 levels in Greenland ice cores stimulated him to connect this interest with his oceanographic interests. The result was a surprising and important calculation. The key was what Broecker later described as a "great conveyor belt" of seawater carrying heat northward. The gross properties of the circulation had been laid out a decade earlier by the GEOSECS survey of radioactive tracers. But it was only now, as

Broecker and others worked through the numbers in enough detail to make crude computational models, that they fully grasped what was happening. The vast mass of water that gradually creeps northward near the surface of the Atlantic is as important in carrying heat as the familiar and visible Gulf Stream. The energy carried to the neighborhood of Iceland was "staggering," Broecker realized—nearly a third as much as the Sun sheds on the entire North Atlantic.[18] If something were to shut down the conveyor, climate would change across much of the Northern Hemisphere. In 1985 Broecker and two colleagues published a paper titled "Does the Ocean-Atmosphere System Have More Than One Stable Mode of Operation?" Their answer was yes: the great conveyor belt could easily shut down.

In one sense this was no discovery. It was an extension of the conjecture that Chamberlin had offered at the start of the century, that the thermohaline circulation, the world-spanning movement of seawater, could shut down if the North Atlantic surface water became less salty (Chapter 1). Few scientific "discoveries" are wholly new. An idea moves from a casual speculation to a discovery when something makes it look truly plausible. Broecker did that by calculating solid numbers. He also pointed out evidence that such a shutdown had actually happened. Geological studies showed that at the end of the last glacial epoch, as the North American ice sheet melted, it had dammed up a huge lake; when this was suddenly released, a colossal flood of fresh water had surged into the ocean. It seemed likely that this had caused a shutdown and cooling. The timing was right: just at the start of the Younger Dryas.

The newly coupled ocean-atmosphere computer models were turned on the question. They found that the melting of a continental ice sheet was not necessary to stop the thermohaline circulation. The balance might be shifted, for example, by additional freshwater rainfall—which might accompany global warming. Bryan and Manabe, working separately, each found that the ocean circulation was so delicately balanced that greenhouse warming might well bring it to a gradual halt. Even at present CO_2 levels, if anything shut it down, it would stay that way. And if the circulation ever did cease, the consequences for climate all around the North Atlantic could be severe.

Broecker was foremost in taking the disagreeable news to the public. In 1987 he wrote that we had been treating the greenhouse effect as a "cocktail hour curiosity," but now "we must view it as a threat to human beings and wildlife." The climate system was a capricious beast, he said, and we were poking it with a sharp stick.[19]

CHAPTER SEVEN

BREAKING INTO POLITICS

Around 1966 Roger Revelle gave a lecture about the Earth's future to students at Harvard University. Among the undergraduates was a senator's son: Albert Gore Jr. The prospect of greenhouse warming came as a shock, Gore later recalled, exploding his childhood assumption that "the Earth is so vast and nature so powerful that nothing we do can have any major or lasting effect on the normal functioning of its natural systems."[1]

By 1981 Al Gore was a representative in Congress and better prepared than anyone to haul climate change from the halls of scientific discussion onto the stage of political controversy. Over the years he had kept abreast of the technical issues as they developed, and he shared the concern about global warming as it grew among scientists. No doubt Gore also saw a political opening. As a champion of environmental issues, he could display leadership in one of the few areas where the newly installed Republican administration's policies disturbed a large majority of voters.

Ronald Reagan had assumed the presidency with an administration that openly scorned environmental worries, global warming included. Many conservatives lumped all environmental concerns together as the rantings of liberals hostile to business, a Trojan horse for the expansion of government regulation and secular values. The recently established National Climate Program Office found itself serving, as an observer put it, as "an outpost in enemy territory."[2] The new administration laid plans to slash funding for CO_2 studies in particular, deeming such research unnecessary.

Gore and a few other representatives decided to embarrass the administration with congressional hearings on the proposed cuts. Some newspapers noticed the testimony by persuasive scientists like Revelle and Schneider. As an aide close to the political process put it, "The popular media is the most potent way of convincing a member of Congress that he should pay attention to scientific issues." Politicians did not read scientific journals; they relied on the press as the "prime detector of the public's fears."[3]

When it came to deciding which scientific developments were "news," American journalists tended to take their cues from the New York Times. The editors of the Times followed the advice of their veteran science writer, Walter Sullivan. A lanky and amiable reporter, Sullivan had frequented meetings of geophysicists ever since the International Geophysical Year of 1957–1958, cultivating a number of trusted advisers. On the subject of climate he began listening to scientists like Schneider, Hansen, and a few others who were alarmed about global warming and determined to attract attention to the issue.

In 1981, for example, Hansen sent Sullivan a scientific report he was about to publish, which announced that the planet was getting noticeably warmer (Chapter 6). The Times put it on page 1 and followed up with an editorial declaring that although the greenhouse effect was "still too uncertain to warrant total alteration of energy policy," it was "no longer unimaginable" that a radical policy change might become necessary.[4] The Department of Energy responded by reneging on funding it had promised Hansen, and he had to lay off five people from his institute.

Everything connected with atmospheric change had become politically sensitive. Thus, scientists were reporting that "acid rain" was ravaging forests (and even the paint on houses) thousands of miles downwind from smokestacks that emitted sulfates. When environmentalists demanded restrictions, the coal industry counterattacked with its own scientists and advertising to promote an image of benign economic progress that could never cause long-range damage.

A still more vehement technical controversy erupted on Halloween 1983, when a group of atmospheric scientists held a carefully orchestrated press conference to announce an unexpected climate risk. Some of them had applied computer models of the effects of aerosols to the aftermath of a nuclear war. They calculated that the smoke from burning cities could bring on a worldwide "nuclear winter"—cold and darkness that might threaten the very survival of humankind. In the fore was Carl Sagan, whose fame—much more as an astronomy popularizer than as an atmospheric scientist—attracted television coverage. Would launching a nuclear attack, even if the other side never fired back, be literally suicidal? So maintained Sagan and his allies, frankly intending to pressure governments to reduce their nuclear arsenals. Meanwhile, their work added another layer to public imagination of global climate catastrophe.

The Reagan administration heaped scorn on these critics of its military buildup. Other scientists questioned the calculations; later the group admitted that war might bring only a serious but survivable "nuclear autumn," but at first most commentators either swallowed the "winter" prediction whole or scoffed at it. The shouts of partisan combat increasingly drowned out reasoned public discussion. Computer calculations of the effects of aerosols had become inescapably entangled in national politics. If you knew a person's views on nuclear disarmament, you could probably guess what the person thought of the nuclear winter prediction. And if you knew a person's views on government regulation, you could probably guess what the person thought of predictions about acid rain—and about global warming.

Most scientists preferred to stand aside from all such debates. In the United States the one accepted authority for providing policy advice was the National Academy of Sciences. In 1980 Congress had asked the Academy to carry out a comprehensive study on the impacts of rising CO_2. The Academy appointed a panel of leading experts, and in 1983, following a sustained effort to work out a consensus, the panel reported. The scientists said they were "deeply concerned" about the environmental changes that they expected a temperature rise would bring. And they pointed out that "we may get into trouble in ways that we have barely imagined"—for example, if warming released methane from seabed sediments. But on the whole, the panel was cautiously reassuring. They said the warming would probably not be very severe. And after all, a degree or two of temperature change was something people in the past had managed to get through well enough. The Academy's chief recommendation was that before doing anything, the government should fund vigilant monitoring and other studies. All they asked was to spend more money on research, explicitly advising against any immediate policy change such as restricting use of fossil fuels.[5]

Three days later the Environmental Protection Agency released a report of its own on the greenhouse effect. The science was mostly the same, but the tone of the EPA's conclusions was more anxious. A ban on fossil fuels seemed out of the question on both economic and political grounds, so the panel saw no practicable way to prevent a temperature rise. It could be a big rise, within a few decades, with potentially "catastrophic" consequences. This EPA report was the first time a federal agency declared that global warming was, as a reporter put it, "not a theoretical problem but a threat whose effects will be felt within a few years."[6]

Administration officials dismissed the EPA report as alarmist, pointing to the more reassuring Academy report. Here was a tale of battling views, just what journalists needed to make a lively story. It spread through the newspapers and even got onto national television. The controversy, piled on top of congressional hearings and the efforts of scientist publicists, alerted a sizable fraction of citizens and politicians to the prediction that was shared by both reports. It was official: global warming might well be coming. Climate scientists found themselves in demand to give tutorials to journalists, government agency officials, and even groups of senators, who would sit obediently for hours of lecturing on greenhouse gases and computer models.

If global warming was coming, what would be the consequences? Some things seemed quite certain on basic physical principles, confirmed by computer models. Science reporters took care that the public, if they read about the topic at all, understood that an *average* warming of 3° did not mean that the thermometer would be exactly 3° higher, everywhere, every day. In some regions the weather might not change much. Other regions would suffer unprecedented heat waves. Less obviously, the air would hold more moisture, so the water cycle would intensify. Some regions would have worse droughts as the moisture evaporated, others would have more storms and floods as the water came back to earth.

And the sea level could rise. Researchers had not been able to dismiss worries about a collapse of the West Antarctic Ice Sheet. Too little was known about the way ice behaved for experts to agree on any firm conclusion. Some studies saw the possibility of a collapse and sea-level rise of 2 to 3 meters (6 to 10 feet) by the year 2100. Most experts disagreed, calculating that this could not possibly happen so soon. Yet a gradual discharge of ice over the following centuries was possible, and that could place a heavy burden on human society.

A smaller but significant rise of sea level within the coming century was expected for another reason. Amid all the talk of melting ice, for decades nobody seems to have given a thought to another simple effect: water expands when heated. Eventually, in 1982 two groups separately calculated that the global warming observed since the mid-nineteenth century must have raised the sea level significantly by plain thermal expansion of the upper ocean layers. There had indeed been a rise of 10 centimeters or so (several times faster than the average change in earlier millennia). Thermal expansion could not account for all of that, and the scientists figured the rest came from melting glaciers—many of the world's small mountain glaciers were in fact shrinking. By the late twenty-first century, they warned, rising tides would erode shorelines back hundreds of feet. Salt water would invade estuaries. Entire populations would flee from storm surges.

So it was widely agreed that global warming could be a threat, and that the proper response was to study it. After all, the late twenty-first century was so far away! Weary of the issue and distracted by more urgent matters, the media and the public turned their attention elsewhere. Funding for the environmental sciences was not drastically cut after all, but neither was it expanded. During the 1980s the U.S. government spent barely $50 million per year for research directly focused on climate change. It was a trivial sum compared with many other research programs. Other nations did not take up the slack. Through Western Europe to the Soviet Union, governments scarcely increased their spending on science in general and climate science in particular.

Of the money that government agencies did make available, they diverted part into a new type of study: the social and economic impacts of climate change. What would it mean for agriculture, forestry, the spread of tropical diseases, water supply systems, and so forth? A new field joined the many others related to climate science: "impact studies." Forecasting impacts required a broadly multidisciplinary approach. Climate scientists began to interact with experts in agriculture, epidemiology, water systems engineering, and many other fields.

For the study of climate change itself, researchers in different specialties increasingly needed to communicate with one another.

They no longer spoke of studying "climates" in the old sense of regional weather patterns, but rather of "the climate system" of the whole planet—everything from minerals to microbes, not to mention the rapidly increasing influences of human activities. Specialists in the quirks of the stratosphere, volcanoes, ocean chemistry, ecosystems, and so forth found themselves sharing the same funding agencies, institutions, and even university buildings. Meanwhile, scientific meetings devoted to one or another interdisciplinary topic also grew common.

Such collaboration was increasing in all the sciences. As research problems spanned ever more complexities, scientists with different types of expertise needed to exchange ideas and data, or work together directly. Before 1940 nearly every paper published on climate had a single author. By the 1980s the majority of papers had more than one author, and papers with seven or eight authors were no longer surprising.

None of this entirely solved the problem of fragmentation. The more the research enterprise grew, the more narrowly scientists were driven to specialize. Meanwhile, the imperatives of administration maintained boundaries between academic disciplines and divided the government agencies and organizations that supported them. Most scientific papers, now as in the past, were published in journals dedicated to a particular field, like the meteorologists' *Journal of the Atmospheric Sciences* or the paleontologists' *Quaternary Research.* Every scientist read *Science* and *Nature,* however. These comprehensive journals competed with one another for the most important papers in every scientific field, including those connected with climate change. Both journals also published expert reviews and staff commentaries that helped keep scientists up to date on developments outside their own field.

The dispersion of research funding among several agencies of the U.S. government was partly solved by the Earth System Sciences Committee, which NASA set up in 1983. The committee members struck bargains among agencies and disciplines, forging a common front. After fighting out their differences among themselves, the members agreed on a short list of top-priority programs and put the weight of their joint prestige behind it. The administration's budget makers and Congress, pleased to see a well-coordinated effort, opened their pockets. Meanwhile, the committee worked out cooperation with scientists around the world. Climate research was finally getting its own international organization, strong enough to win support from individual nations.

The hundreds of scientists and government officials who assembled at the World Climate Conference held in Geneva in 1979 (Chapter 5) called for an international structure set up specifically for climate research. The government representatives who constituted the WMO and scientific leaders who constituted the ICSU took that advice and joined to launch the World Climate Research Program (WCRP). It took over the portion of the old Global Atmospheric Research Program that had been concerned with climate change, including a small staff in Geneva and an independent scientific planning committee. The WCRP had various branches, each known mainly by an acronym. For example, the International Satellite Cloud Climatology Project (ISCCP) collected streams of raw data from the weather satellites of several nations and fed the data through a variety of government and university groups for processing, analysis, and archiving. These U.N.–sponsored efforts were only one strand, although the central one, in a tangle of national, bilateral, and multinational climate initiatives. Countless organizations were now seeking to be part of the action.

Of course, none of this proliferation was really the work of abstract entities. It was brought about by a few scientists and officials dedicated to international and environmental interests. They blurred the distinction between governmental and private initiatives as they organized international meetings. The indispensable man was Bert Bolin, who kept busy chairing meetings, editing reports, and promoting the establishment of panels. Along with his outstanding personal abilities as a scientist, executive, and diplomat, Bolin benefited from his position at the University of Stockholm in Sweden, traditionally neutral territory.

The most important initiative was a series of invitational meetings for meteorologists held at Villach, Austria, through the 1980s. The 1985 Villach conference was a turning point. Ramanathan had just announced that the greenhouse effect due to CO_2 would be doubled by the rise of trace gases like methane. So global warming was not just a problem for the late twenty-first century; it might become serious within the scientists' own lifetimes. The assembled experts arrived at an international consensus: "In the first half of the next century a rise of global mean temperature could occur which is greater than any in man's history." And for the first time, a group of climate scientists went beyond the usual call for more research to take a more activist stance. Governments should act, and soon. "While some warming of climate now appears inevitable due to past actions," they declared, "the rate and degree of future warming could be profoundly affected by governmental policies."[7]

Villach and other international meetings, along with similar consensus-building studies on climate change carried out by national bodies such as the U.S. National Academy of Sciences, crystallized a set of beliefs and attitudes among climate scientists. As one science writer reported after a series of interviews, "By the second half of the 1980s, many experts were frantic to persuade the world of what was about to happen."[8]

Human motivation is seldom simple, and behind the emotional commitment of scientists lay more than dry evaluation of data. Adding to their concern about the changing climate was the normal tendency of people to perceive their own fields as important (with the corollary that funds should be generously awarded for their work and for their students and colleagues). The scientists found allies among administrators in national and international bureau-

cracies. Warnings of future dangers reinforced the normal inclination of officials to extol the importance of their areas of responsibility as they sought bigger budgets and broader powers. Whenever
there is evidence that something needs to be done, those who stand
to profit from the doing will be especially quick to accept the evidence and argue for policy changes. The few politicians who joined,
like Gore, also hoped that personal convictions would fit with opportunities for career advancement.

The only reliable way to sort through the human motives and
determine what policy action was really needed was to seek rigor-
ous scientific conclusions. Although a few scientists and officials
tentatively proposed policy changes, many more were pushing for
still larger international research projects. The WCRP was all to
the good, but it stuck too narrowly to meteorology. Around 1983
various organizations collaborated under ICSU to draw together
all the geophysical and biological sciences in the International
Geosphere-Biosphere Program (IGBP). Starting up in 1986, the
IGBP built a large structure of committees, panels, and working
groups that fostered interdisciplinary connections. The drawback,
as Schneider pointed out, was the unavoidable fact that "an IGBP
should be in the business of measuring or modeling everything at
once from the mantle of the Earth to the center of the Sun!"[9]

Research did result in a big policy breakthrough in the late 1980s,
although not for climate. Ever since the spray-can controversy of
the mid-1970s, scientists had worried about the destruction of ozone
in the stratosphere. In 1985 this led twenty nations to sign the
Vienna Convention for the Protection of the Ozone Layer. The doc-
ument was only a toothless expression of hopes, but it established
a framework, which became useful almost at once. In 1985 a British group announced their discovery of a "hole" in the ozone layer
over Antarctica. The apparent culprit was again CFCs, banned from
American spray cans but still widely produced around the world
for a variety of purposes. Controversy inevitably broke out. Indus-

trial groups denied their products could be risky and declared that changing them would be an insufferable economic drag. Reagan administration officials reflexively backed the corporations against all environmentalist claims.

The denials were short-lived. Within two years, new theories of how the chemicals could destroy ozone, confirmed by daring flights over Antarctica, convinced the experts. The immediate threat of increased skin cancers and other damage to people and biological systems shocked officials. Meanwhile, magazine and television images of the ominous map of ozone loss carried the message to the public. Most people lumped together all forms of potential atmospheric harms, mingling the threat to the ozone with greenhouse gases, smog, acid rain, and so forth. Politicians were forced to respond. In the epochal 1987 Montreal Protocol of the Vienna Convention, the world's governments formally pledged to restrict emission of specific ozone-damaging chemicals.

This was not the first international agreement to restrict pollution in response to scientific advice. In 1979, for example, the nations of Western Europe had adopted a convention to address acid rain, pledging to study their sulfate emissions and impose limits on them. The Montreal Protocol set an even stricter standard for international cooperation and national self-restraint. Over the following decade, it had great success in reducing emissions of CFCs. Although essential for protecting the ozone layer, the suppression of these chemicals was not much help for climate. Some of the chemicals that industry substituted for CFCs were themselves greenhouse gases.

At the same time, many environmentalists hoped that the precedent set by the Montreal Protocol could show the way to restrictions on greenhouse gases. It turned out that market-oriented mechanisms could be devised to regulate CFCs much more cheaply than the industries had predicted. As with acid rain and most other pollutants, over the long run the nations that imposed regulations enjoyed a net economic benefit.

Governments followed up the Montreal success the next year, 1988, with the World Conference on the Changing Atmosphere: Implications for Global Security, nicknamed the Toronto Conference. The planning came out of the workshops initiated by the 1985 Villach conference. Like Villach, Toronto was a meeting by invitation of scientist experts—not official government representatives, who would have had a much harder time reaching a consensus. In the conference report, for the first time a group of prestigious scientists called on the world's governments to set strict, specific targets for reducing greenhouse gas emissions. That was the Montreal Protocol model: set targets internationally and let governments come up with their own policies to meet them. By 2005, said the experts, emissions ought to be pushed some 20 percent below the 1988 level. "If we choose to take on this challenge," one of the scientists declared, "it appears that we can slow the rate of change substantially, giving us time to develop mechanisms so that the cost to society and the damage to ecosystems can be minimized. We could alternatively close our eyes, hope for the best, and pay the cost when the bill comes due."[10]

Up to this point global warming had been mostly below the threshold of public attention. The reports that the 1980s were the hottest years on record had barely made it into the inside pages of newspapers. A majority of people were not even aware of the problem. Those who had heard about global warming saw it mostly as something that the next generation might or might not need to worry about. Yet a shift of views had been prepared by the ozone hole, acid rain, and other atmospheric pollution stories, by a decade of agitation on these and many other environmental issues, and by the slow turn of scientific opinion toward strong concern about climate change. To ignite the worries, only a match was needed. This is often the case for matters of intellectual concern. No matter how much pressure builds up among concerned experts, some trigger is needed to produce an explosion of public attention.

The break came in the summer of 1988. A series of heat waves and droughts, the worst since the Dust Bowl of the 1930s, devastated many regions of the United States. Cover articles in newsmagazines, lead stories on television news programs, and countless newspaper columns offered dramatic images of parched farmlands, sweltering cities, a "super hurricane," and the worst forest fires of the century. Reporters asked: were all these caused by the greenhouse effect? Scientists knew that no individual weather event could be traced to global warming. But simply from endless repetition of the question, many people became half convinced that our pollution was indeed to blame for it all.

In the middle of this Hansen raised the stakes with deliberate intent. By arrangement with Senator Timothy Wirth, he testified at a congressional hearing in late June, deliberately choosing the summer, hardly a normal time for politicians who sought attention. Outside the room the temperature that day reached a record high. Inside, Hansen said he could state "with 99 percent confidence" that there was a long-term warming trend under way, and he strongly suspected that the greenhouse effect was to blame. Talking with reporters afterward, Hansen said it was time to "stop waffling, and say that the evidence is pretty strong that the greenhouse effect is here."[11] As the heat waves and drought continued, reporters descended unexpectedly on the Toronto conference and prominently reported its alarming conclusions. The story was no longer a scientific abstraction about an atmospheric phenomenon: it was about a present danger to everyone, from elderly people struck down by heat to the owners of beach houses. Images of blasted crops and burning forests seemed to be a warning signal, a visible preview of what the future might hold.

The media coverage was so extensive that, according to a 1989 poll, 79 percent of Americans recalled having heard or read about the greenhouse effect. This was a big jump from the 38 percent who had heard about it in 1981, and an extraordinarily high level of public awareness for any scientific phenomenon. Most of these citizens thought they would live to experience climate changes.

The environmental movement, which had taken only an occasional interest in climate change, now took it up as a main cause. Groups that had other reasons for preserving tropical forests, promoting energy conservation, slowing population growth, or reducing air pollution could make common cause as they offered their various ways to reduce emissions of CO_2. Adding their voices to the chorus were people who looked for arguments to weaken the prestige of large corporations. For better or worse, global warming became firmly identified as a "green" issue.

In the long perspective, it was an extraordinary novelty that such a thing became a political question at all. Global warming was invisible, no more than a possibility, and not even a current possibility but something predicted to emerge only after decades or more. The prediction was based on complex reasoning and data that only a scientist could understand. It was a remarkable advance for humanity that such a thing could be a subject of widespread discussion. Discourse had slowly been growing more sophisticated. Not only had science made enormous gains, but the general public had become better educated.

Politicians at the highest level began to pay attention to greenhouse gases. The United Kingdom's prime minister, Margaret Thatcher, had once worked as a chemist and fully understood what scientists told her. In 1988 she became the first major political leader to take a determined stand on global warming, calling it a key issue and dedicating new funds to research. Attention from the politically powerful "Greens" in Germany and elsewhere in continental Europe added weight. What had begun as a research puzzle had become a serious international issue.

Negotiations about regulations would be shaped (at least in part) by scientific findings. That was a heavy burden for climate scientists. In the late 1980s climate research topics became far more prominent in the scientific community itself. The volume of publications soared, and there were so many workshops and conferences that nobody could attend more than a fraction of them. This sudden growth of scientific attention was partly due to the step-up in public concern: anyone studying the topic would now get a better hearing when requesting funds, recruiting students, and publishing. In turn, the new emphasis on climate research was reflected by science journalists, who took their cues from the scientific community and passed its views on to the public.

Yet there were still only a few hundred people in the world who devoted themselves full-time to the study of climate change. And they were still dispersed among many fields. Somehow their expertise had to be assembled to answer the urgent policy questions. The reports issued by brief conferences could not command thorough respect, and they did not commit any particular group to follow up. Programs like the IGBP were designed only to promote an array of research projects. Some new kind of institution was needed.

Conservatives and skeptics in the U.S. administration might have been expected to oppose the creation of a prestigious body to study climate change. But they distrusted still more the system of international panels of independent scientists that had been driving the issue. If the process continued in the same fashion, the skeptics warned, future groups might make radical environmentalist pronouncements. Better to form a new system under the control of government representatives. Besides, a complex and lengthy study process would restrain any move to take concrete steps to limit emissions.

Responding to all these pressures, in 1988 the WMO and other United Nations environmental agencies created the Intergovernmental Panel on Climate Change (IPCC). Unlike earlier climate panels, the IPCC was composed largely of representatives of the world's governments—people who had strong links to national laboratories, meteorological offices, science agencies, and so forth. It was neither a strictly scientific nor a strictly political body, but a unique hybrid.

Most people were scarcely aware of the central factor in these international initiatives: the worldwide advance of democracy. It is too easy to overlook the obvious fact that international organizations are governed in a democratic fashion, with vigorous free debate and votes in councils. If we tried to make a diagram of the organizations that dealt with climate change, we would not draw an authoritarian tree of hierarchical command, but a spaghetti tangle of cross-linked, quasi-independent committees. Often decisions were made, as they were in the IPCC, by a negotiated consensus in a spirit of equality, mutual accommodation, and commitment to the community process (a seldom-appreciated part of the democratic culture).

It is an important fact that all such international organizations have been created primarily by governments that felt comfortable with such mechanisms at home, that is, democratic governments. Happily, the number of nations under democratic governance increased dramatically during the twentieth century, and by the end of the century they were predominant. Therefore, democratically based international institutions proliferated, exerting an ever stronger influence in world affairs.[12] This was seen in all areas of human endeavor, but it often came first in science, internationally minded since its origins. The democratization of international politics was the scarcely noticed foundation on which the IPCC and its fellow institutions took their stand.

It worked both ways. The international organization of climate studies helped fulfill some of the hopes of those who, in the aftermath of the Second World War, had worked to build an open and cooperative world order. If the IPCC was the outstanding example, in other areas ranging from disease control to fisheries, panels of scientists were becoming a new voice in world affairs. Independent of nationalities, they wielded increasing power by claiming dominion over facts about the actual state of the world—thus shaping perceptions of reality itself. Such a transnational scientific influence on policy matched dreams held by liberals since the nineteenth century. It awoke corresponding suspicions among the enemies of liberalism.

CHAPTER EIGHT

SPEAKING SCIENCE TO POWER

When a nation's leaders face a hostile foreign power or an economic crisis, they will act. However imperfect the policy makers' information about the risk, they know that to do nothing is itself a choice among actions, and seldom the best choice. The risk of climate change brought a different response. Here was an altogether unfamiliar type of danger, a danger that would have remained hidden but for a few scientists whose names hardly anybody knew, a danger described only by computer programs that hardly anybody understood.

For decades, policy makers had responded simply by spending money on research (if never enough). It was up to the world's scientists to declare the nature of the danger, while holding to the strict rules of evidence and reasoning that made them scientists in the first place. And no declaration would have force unless it emerged from the clumsy mechanism of the Intergovernmental Panel on Climate Change, imposed by governments that had no wish to hear bad news.

Unlike all earlier climate conferences and panels, the IPCC was an official intergovernmental effort, and with virtually all the world's climate scientists and governments participating, it quickly established itself as the principal source of advice to policy makers. The key to the panel's success was building its bureaucratic structure out of familiar components—the flexible, freewheeling groups that scientists had perfected over centuries of research and controversy. Independent task forces addressed each of the chief scientific issues. Experts reviewed the latest research publications and drafted reports. In informal workshops, additional experts debated every detail for days on end. Meanwhile, individuals argued with one another in letters and, increasingly, e-mails. All through 1989, 170 scientists in a dozen workshops labored to craft statements that none could fault on scientific grounds. These reports went through yet another review that gathered comments from virtually every significant climate expert in the world (and many of the insignificant ones). In the end, the scientists found it easier than they had expected to achieve a consensus. One reason was the judicious chairmanship of Bert Bolin. His soft-spoken diplomacy served a steely determination to spell out exactly what was known, and what was not known, about the risk of global warming.

The scientists' findings then had to be endorsed unanimously by the official government delegates, many of whom were not scientists at all. The most passionate in arguing for a strong statement were representatives of small island nations, who feared that rising seas would eventually erase them from the map. Far more powerful were the oil and coal industries, represented by governments of nations living off fossil fuels, like Saudi Arabia. The negotiations were intense, and only Bolin's skills and the fear of an embarrassing collapse pushed people through the grueling sessions to grudging agreement. The final consensus statements, negotiated word by word, were highly qualified and cautious. This was not mainstream science so much as lowest-common-denominator science. But when the IPCC finally announced its conclusions, every word had solid credibility.

Issued in 1990, the first IPCC report concluded that the world had indeed been warming. The report predicted (correctly) that it would take another decade before scientists could say with any confidence whether the warming was caused by natural processes or by humanity's greenhouse gas emissions. Still, the panel thought it likely that human emissions would cause global warming, perhaps a few degrees by the mid-twenty-first century—which still seemed far away. The report contained no exciting or surprising "news" and got hardly any press coverage.

A closer look at the IPCC report would have found more food for thought. When the panel said the future was uncertain, that did not mean the risk to climate should be ignored. The experts had pointed out economically sound ways to get a start on reducing the risk, a broad hint that governments should begin to act.

Other groups, ranging from government agencies to environmental organizations, called more vigorously for action. A 1991 National Academy of Sciences report listed no fewer than fifty-eight policies that could mitigate greenhouse warming. Some were "no-regrets" policies, so practical that they would benefit the economy even if there was no climate problem. For example, governments might promote improvements in the efficiency of commercial lighting, home heating, and trucks. Or they could reduce the enormous hidden and overt subsidies that encouraged wasteful consumption of fossil fuels. Some policies would carry a modest cost that would be compensated for by valuable social benefits. Why not, for example, devise ways to reduce car commuting time, and reforest overgrazed wastelands? Some ideas were too expensive, but might become practical in the future if technology was driven forward by the regulation or taxation of greenhouse gas emissions, or by plain desperation. For example, it might someday make sense to capture CO_2 from a power plant as it burned fuel and sequester the gas underground. And some proposals were visionary. Couldn't we shield ourselves by spreading sulfate particles in the stratosphere, or launch flotillas of mirrors into space to reflect sunlight away from the Earth?

None of this attracted much public interest. After the spate of global warming stories in 1988, media attention had inevitably declined as more normal weather set in and editors looked for fresher topics. Environmental organizations continued to argue for restrictions on emissions with sporadic lobbying and advertising efforts. They were opposed, and greatly outspent, by a professional public relations campaign. In the forefront was the Global Climate Coalition, funded by dozens of major corporations in the petroleum, automotive, and other industries connected with fossil fuels. With slick publications and videos sent wholesale to journalists, along with lobbying in Washington and at international meetings, the coalition did much to persuade policy makers and citizens that there was no need to worry about climate change. Corporations and wealthy individuals also channeled money to individual scientists and a dozen conservative think tanks that insisted there was no global warming problem. This effort resembled the glib scientific criticism and persuasive advertising that other industries had used to attack scientists' warnings about tobacco smoke, ozone depletion, and acid rain. Although each of those campaigns had been discredited after a decade or two, fair-minded people were ready to listen to the global-warming skeptics.

Given the real scientific uncertainties, it seemed reasonable that government regulation to reduce emissions was unnecessary or at least premature. Conservatives argued that a strong economy (which they presumed meant one with the least possible government regulation of industry) would offer the best insurance if problems ever did arise. Activists replied that action to retard the damage should begin as soon as possible. They argued hardest for policy changes that they wanted for other reasons, such as protecting tropical forests or subsidizing public transport to reduce urban smog.

Surveys of climate experts in the early 1990s found that a majority believed that significant global warming was likely. When asked to rank their certainty about this on a scale from one to ten, most picked a number near the middle. Only a few were fairly confident either that global warming would not come or that it would. Nevertheless, roughly two-thirds of the climate experts felt that people ought to start taking steps to lessen the danger, just in case.[1]

When corporations and conservatives denied any need to act against global warming, the first issue they had to confront was the actual temperature rise. News media now reported fairly prominently the annual summaries of the planet's temperature issued by the New York and East Anglia groups. The year 1988 had been a record breaker.

A few scientists claimed that there had actually never been any global warming at all, that the statistics showing a strong rise since the nineteenth century were an illusion. The most prominent of these was S. Fred Singer, who had retired in 1989 from a distinguished career managing government programs in weather satellites and other technical enterprises. He founded an environmental policy group supported by conservative foundations. Singer and others insisted that climatologists had overlooked the "urban heat island effect": the fact that cities are usually warmer than the countryside around them. They argued that by measuring temperatures in the world's growing cities, climatologists had given a false picture of global warming. In fact (as Singer knew), every expert since Callendar had taken this effect into account in their calculations. And anyway (as Singer knew), temperatures were also rising over the unpopulated oceans and Arctic wastes.

Other skeptics conceded that global temperatures really had risen. Their argument was that the rise was merely a normal, random fluctuation like many others in history, perhaps a "recovery" from the Little Ice Age. Besides, in the early 1990s the average global temperature declined slightly. The critics scoffed at the computer models that predicted a future of greenhouse warming.

The modelers admitted they still had much to learn. One galling problem was that different models gave different predictions for how the climate would change in a given locality. For example, although models consistently projected that severe droughts would afflict the American Southwest, for the Southeast some models saw less rain while others saw more. Turning to the past, for the known increase of greenhouse gases from the turn of the century to the 1990s, most models calculated a 1°C global rise. The actual rise, however, had been at most half that. Critics remarked that the models had to share some flaw, something they collectively overlooked. Nevertheless, most experts felt the GCMs were on the right track. Take any kind of model that gave a reasonably realistic picture of the present climate, feed in an increase of CO_2, and the invariable result was some kind of global warming.

Although only a few respected scientists insisted that computer predictions had no value, a powerful apparatus amplified their voices to a shout. For example, brief reports published by the conservative George C. Marshall Institute assembled a well-argued scientific skepticism. They also insisted that regulation of greenhouse emissions would be "extraordinarily costly to the U.S. economy."[2] The authors of these pamphlets were anonymous. This was far from the standard practiced by the peer-reviewed scientific journals. Established journals like *Science, Nature,* the *Journal of Geophysical Research,* and dozens of others were now carrying hundreds of climate articles a year, each representing countless hours of meticulous research. Scarcely one of these papers disputed the fundamentals of greenhouse warming. The denials were found mostly in publications funded by industrial groups and conservative institutions, or in right-wing venues like the *Wall Street Journal*'s editorial page. Some climate experts publicly denounced the skeptics' reports as "junk science."[3] Open conflict broke out in acrimonious personal exchanges.

To science journalists and their editors, the controversy made excellent story material. Outside the more conservative media, many reporters believed global warming was likely, and they reached for attention with dire predictions of devastating droughts, ferocious storms, waves attacking drowned coastlines.[4] But they also reported on the people who called that nonsense. Conflict always makes a good story, and many reporters presented climate science as a dispute between two equal and diametrically opposed teams. They often sought an artificial balance by pitting "pro" against "anti" scientists, one to one. A handful of persuasive skeptics got quoted again and again, including some who had done little science of note. It was hard to recall that there was a consensus view shared by a very large majority of experts: the IPCC's modest 1990 conclusion that global warming was far from certain, but it was a truly serious possibility.

Governments in Western Europe and Japan did grasp that they should consider serious policy changes, although they were satisfied with studying how to go about it. Some U.S. agencies also wanted to get to work restraining greenhouse gases. But President George H. W. Bush's administration, its beliefs shaped by industry lobbyists, denied any need to address climate change. A White House memorandum, inadvertently released in 1990, endorsed the conservatives' view that the best way to address concern about global warming was "to raise the many uncertainties."[5]

The U.S. government's overt rejection of the IPCC's conclusions became an embarrassment in 1992. World leaders were preparing their grandest meeting ever, the "Earth Summit" in Rio de Janeiro (officially the United Nations Conference on Environment and Development). The great majority of governments called for negotiating mandatory limits on greenhouse gas emissions. But no negotiation could get far without the United States, the world's premier political, economic, and scientific power—and largest emitter of greenhouse gases. The American administration, attacked by its closest foreign friends as an irresponsible polluter, showed some flexibility. Diplomats papered over disagreements to produce a compromise that included targets for reduced emissions. An agreement (officially the "United Nations Framework Convention on

Climate Change") was signed at Rio by more than 150 govern-ments. The agreement's evasions and ambiguities left so many loop-holes that policy makers could avoid meaningful action. Indeed, scarcely any nation did more than bestow some research funds and lip service. Still, the Framework Convention did establish basic principles for what governments ought to do eventually, and it mapped a path for further negotiations.

The creation of the IPCC had established a cyclic international process. Roughly twice a decade, the IPCC would analyze the most recent peer-reviewed research and issue a consensus statement about the prospects for climate change. That would lay a foundation for international negotiations, which would establish guidelines for in-dividual national policies. Further moves would await the results of further research. Thus, after governments responded to the Rio convention (with inaction, as it happened), it was the scientists' turn. They pursued research problems as usual, published the re-sults in journals as usual, and discussed technicalities in confer-ences as usual, but to officialdom that was all in preparation for the next IPCC report, scheduled for 1995.

The computer modelers pushed ahead without any big break-throughs, a steady, slogging job. One group would find a better way to represent the formation of clouds, another would come up with a more efficient way to calculate winds, and once or twice a decade each group would convince some agency to buy them a supercom-puter many times faster than the last. Thanks to the fabulous ex-pansion of processing power, groups could now routinely couple ocean and atmosphere models to represent the climate system as a whole in elaborate detail. Their work was greatly helped by the fact that they could set their results against a uniform body of world-wide data. Specially designed satellite instruments were monitoring incoming and outgoing radiation, cloud cover, and other essential parameters—showing, for example, where clouds brought warming and where they brought cooling.

Most of all, the modelers had to deal with the growing recognition that global climate change was not a matter of CO_2 alone. Besides other greenhouse gases like methane, human activity was now providing at least a quarter of the atmosphere's dust and chemical haze. These agents influenced radiation directly or through their effects on clouds. In particular, as researchers in various fields worked up observations and calculations, they came to agree that sulfate aerosols brought significant cooling.

In 1991 Mount Pinatubo in the Philippines exploded. A mushroom cloud the size of Iowa burst into the stratosphere, depositing some 20 million tons of SO_2. Hansen's group saw an opportunity in this "natural experiment" to test their computer model. Putting the volcano's emissions into their calculations, they boldly predicted roughly half a degree of average global cooling. It would be concentrated in the higher northern latitudes, and would last a couple of years. Exactly such a temporary cooling was in fact observed—this was the pause in record-setting heat waves that encouraged global warming skeptics. The success of Hansen's prediction strengthened the modelers' confidence that they were getting a grip on the effects of aerosols.

As one expert drily remarked, "The fact that aerosols have been ignored means that projections may well be grossly in error."[6] Emissions from the "human volcano" were apparently acting like an ongoing Pinatubo. Computer modelers set to work to incorporate this into their GCMs. The results answered the problem that models had calculated a temperature rise double what was observed. It turned out that the modelers had gotten the effect of the rise of the CO_2 level about right. But now that they could figure in correctly the cooling effect of the rise of pollution, they found that it had counteracted part of the expected greenhouse warming. In 1995 improved GCMs constructed at three different centers (the Lawrence Livermore National Laboratory in California, the Hadley Center for Climate Prediction and Research in the United Kingdom, and the Max Planck Institute for Meteorology in Germany) all succeeded in reproducing the actual twentieth-century temperature rise.

Other data further supported the models' validity. Every calculation back to Arrhenius had found, on elementary principles, that the greenhouse effect would act most strongly at night (when the planet loses heat most rapidly into space). Statistics did show that it was especially at night that the world was warmer. Moreover, Arrhenius and everyone since had calculated that the Arctic would warm up faster than other parts of the globe (as the melting of snow and ice exposed dark soil and water). The effect was glaringly obvious to scientists, and not just from temperature records, for they observed trees advancing into high mountain meadows in Sweden and a thinning of the Arctic Ocean ice pack. Pursuing this in a more sophisticated way, GCMs predicted that greenhouse gases would cause a specific geographical pattern of temperature change, different from the pattern that would result from other influences such as variations in solar energy. The weather data showed a rough resemblance to the greenhouse-effect pattern and nothing else. Scientists called it the long-sought "fingerprint" of the greenhouse effect.

These results exerted a powerful influence as the IPCC finished up work on its next report. After another exhaustive round of workshops, draft reports, formal critiques, e-mail wrangling, and consultation among some four hundred experts plus more than a hundred official national representatives, not to mention lobbying by every variety of industrial and environmental interest, in 1995 the IPCC presented its updated conclusions. The report's single widely quoted sentence said, "The balance of evidence suggests that there is a discernible human influence on global climate."[7] The weaselly wording showed the strain of political compromises that had watered down the original draft, but nobody missed the message.

This second IPCC report, following a tradition now well established, reckoned the consequences of a doubled CO_2 level. Just when that would come would depend on the future of the global economy and its emissions. The result, the panel reported, would be a temperature rise probably somewhere between 1.5° and 4.5°C. That was exactly the range of numbers announced by the first IPCC report and other groups ever since 1979, when the Charney committee had published its rough guess (Chapter 5). Since then computer modeling had made tremendous progress. But the meaning of these numbers had always been hazy. The new numbers actually ran as high as 5.5°C. They had been calculated separately for a half dozen different economic and political scenarios, ranging from low global population growth and stringent emission controls to heedless industrial expansion without any controls at all. The authors of the 1995 IPCC report decided to stick with the familiar old numbers, rather than give critics an opening to cry inconsistency. The meaning of the numbers had invisibly changed: experts had grown more confident that this was a plausible range of temperatures for the most likely choices for our civilization's future growth. It was a striking demonstration of how the IPCC process deliberately mingled science and policy issues until they could scarcely be disentangled.

The somnolent public debate revived briefly in 1995 on the news that the IPCC had agreed that human emissions might well be making the world warmer. As the first formal declaration by the assembled experts of the world, this was front-page news everywhere, immediately recognized as a landmark. "It's official," as *Science* magazine put it—the "first glimmer of greenhouse warming" had been seen.[8] Better still for journalists, the report stirred up a nasty controversy when critics cast doubt on the personal integrity of some IPCC scientists.

National governments felt pressed to respond. In the United States, after Bill Clinton took office as president in 1993, Vice President Al Gore and others persuaded him to commit the nation formally to the Rio target for reducing greenhouse gases. But Gore's views did not reach far in Washington's politics. Many powerful conservatives not only scoffed at any research that pointed to any environmental problem, but also held deep suspicions about the United Nations and all international programs. Faced with these powerful and implacable opponents, Clinton declined to spend his limited political capital on an issue that would not become acute during his term in office.

Neither criticism nor official indifference stopped the international process from rolling forward according to schedule. The next conclave, the 1997 U.N. Conference on Climate Change, gathered in Kyoto, Japan. It was a policy-and-media extravaganza attended by nearly six thousand official delegates and thousands more representatives of environmental groups and industry, and, of course, a swarm of reporters. Representatives of the United States proposed that industrial countries should gradually reduce their emissions to 1990 levels. Most other governments, with Western Europe in the lead, demanded more aggressive action. Coal-rich China and most other developing countries, however, demanded exemption from regulation until their economies caught up with the industrialized nations.

The greenhouse debate was now tangled up with intractable problems of fairness and power relations between industrialized and developing countries. The negotiations almost broke down amid frustration and exhaustion. But the IPCC's conclusions could not be brushed aside. Dedicated efforts by many delegates were capped by a dramatic intervention by Gore, who flew to Kyoto on the last day and pushed through a compromise: the Kyoto Protocol. The agreement exempted poor countries for the time being and pledged wealthy countries to cut their emissions significantly by 2010.

National governments were now supposed to incorporate their pledges into concrete policies. To block any chance of that in the

United States, the Global Climate Coalition mounted a multimillion-dollar advertising and lobbying campaign. Conservatives appealed to nationalism by warning that the Kyoto Protocol would turn over the world economy to the unregulated developing countries. They pointed with horror to the specter of a "carbon tax," a levy on CO_2 emissions that would raise gasoline prices—something Americans supposedly would never tolerate (unlike, say, Europeans). If half a dollar was added to the price of gasoline, they exclaimed, it would destroy the nation's economic health.

Even before the Kyoto delegates had assembled, the U.S. Senate had declared by a vote of 95–0 that it would reject a treaty that exempted developing countries. Afterward, the treaty was never submitted to the Senate for ratification. With little debate, American politicians avoided any policy change that might move toward meeting the Kyoto targets. Most other nations took that as an excuse to carry on likewise with business as usual.

So it was again the turn of the scientists. Skeptics continued to offer plausible technical criticisms, stimulating research in several directions. Particularly influential was an argument that the most probable cause of the modern warming trend was not greenhouse gases at all, but a temporary increase in solar activity. In truth, something besides CO_2 and aerosols was needed to make an exact match with the peculiar rise-fall-rise of temperature since the nineteenth century. A solar connection such as the one that Jack Eddy had sketched out sounded increasingly plausible.

A few scientists proposed complex mechanisms for such a connection, hypothesizing subtle effects of cosmic rays or ultraviolet radiation on cloud formation or ozone. It was possible that such tiny influences could indeed make a difference by interfering in the teetering atmospheric wind patterns. Or perhaps the normal physics of feedbacks was enough to amplify a slight change in the total energy of sunlight. (Precise satellite measurements were finding that this did change by a 0.1 percent as the sunspot cycle rose and fell.) Whatever the mechanism, most scientists came to accept that the climate system was so unsteady that minor changes in the Sun's radiation might provoke noticeable shifts. They now thought it likely that the rise in global temperature up to the 1940s, when greenhouse emissions had still been low, had been caused at least in part by a corresponding increase of sunspot activity. And a dip in solar activity had combined with a surge of aerosol pollution to bring the dip in temperature from the 1940s to the 1970s.

That helped to set rough limits on the extent of the Sun's influence. Average sunspot activity did not increase after the 1970s, and cosmic rays likewise showed no long-term trend (both have remained flat to the time of this writing). Yet the global temperature rise that began in the 1970s had continued. Once the aerosols from the Pinatubo eruption washed out of the atmosphere, the rise accelerated at a record-breaking pace.

The significance of the argument that solar variations influenced climate now reversed. If the planet reacted with such extreme sensitivity to minor changes in the radiation arriving from the sun, it had to be sensitive to greenhouse gas interference with the radiation once it entered the atmosphere. A 1994 National Academy of Sciences panel estimated that if solar radiation was now to weaken as much as it had in the Little Ice Age of the seventeenth century, the effect would be offset by only two decades of accumulation of greenhouse gases. As one expert explained, the Little Ice Age "was a mere 'blip' compared with expected future climatic change."[9]

Critics fell back on the argument that predictions of greenhouse warming depended on computer models, which they continued to insist were valueless. They pointed out that the modelers reproduced the present climate only by laboriously "tuning" their GCMs to match the observed data, adjusting a variety of arbitrary parameters. Some accused the modelers of essentially fudging their results.

The most interesting criticism came from a distinguished Massachusetts Institute of Technology meteorologist, Richard Lindzen.

He offered a scenario involving changes in the way moisture would be carried between layers of the atmosphere as the greenhouse gas level rose, perhaps creating more tropical clouds that would shield the planet by reflecting sunlight. Although Lindzen's argument was complex, he said his thinking rested on a simple philosophical conviction: over the long run natural self-regulation must always win out. Lindzen's technical arguments persuaded none of his peers. Cloud measurements were few and difficult, but the bulk of evidence indicated that the way the modelers handled water vapor was roughly correct.

Another criticism from insiders was that models had attempted a crucial test—and failed. Back in the 1970s oceanographers had cooperated in a large-scale project, CLIMAP, that trolled the seven seas for information on conditions at the peak of the last ice age. From measurements of foraminifera shells brought up from the seabeds, they created a map of ocean temperatures roughly 20,000 years ago. The CLIMAP team reported that in the middle of the last ice age, tropical seas had been only slightly cooler than at present. No matter how much modelers fiddled with GCMs with ice sheets spread over the continents, they could not reproduce that temperature structure. Were the GCMs fundamentally incapable of calculating any climate different from the present one?

The answer came in the late 1990s. It turned out that what had been unreliable was not the computer models, but the oceanographers' complex analysis of their data. New types of climate measures indicated that tropical waters had turned significantly colder in the ice age, agreeing reasonably well with the GCMs. The fact that nobody had been able to adjust a model to make it match the CLIMAP team's numbers now took on a very different significance: evidently the computer models rendered actual climate processes so faithfully that they could not be forced to lie.

Many problems remained. Modelers still could not calculate from first principles the interactions among radiation, water droplets, and aerosols, so they used average parameters to derive the effects of cloud cover. And if they did figure out just how particles affected clouds, they would still have to know what aerosols would actually be present. The modelers had only sketchy measurements of the particles and molecules, let alone data on how they varied from place to place and from season to season. Even if the modelers had known all that, they would have been held back by their ignorance of the chemistry of the atmosphere, where the pollutants interacted and changed over time. Worse, oceanographers still had not solved the mystery of how heat is transported up or down from layer to layer in the seas. Until the real processes could be observed and represented with equations, the risk remained that the ocean circulation could change in some way radically different from what the computers calculated. Finally, even a perfect atmospheric GCM, perfectly coupled to a perfect ocean circulation model, would only make a start. For one thing, it would have to be coupled to models of vegetation.

By the mid-1990s scientists had convincing evidence that changes in vegetation really did alter regional climates. At several locations, overgrazed grasslands with dried-out soils had become demonstrably hotter than intact pastures (and the heating would make it all the harder for grass to return). Some rain forests that were deforested showed a measurable decrease in rainfall, since moisture was no longer evaporated back into the air from the leaves of trees—in Brazil, rain fled from the plow. Meanwhile, a scientist pointed out one more of those ideas that seem obvious, but only after they have been said: if warming made forests grow farther north, the dark pines would absorb more sunlight than snowy tundra and add to the warming.

Some scientists stuck by the old view that biological feedbacks were not alarming but reassuring. They held that fertilization from the increased CO_2 in the atmosphere would benefit agriculture and forestry so much that it would make up for any damage from climate change. Studies found that for the planet as a whole, biomass was indeed absorbing more CO_2 overall than in earlier decades. The consequences were not straightforward, however. The extra CO_2 might benefit invasive weeds and insect pests more than desirable crops. In any case, as the level of the gas continued to rise, plants would reach a point (nobody could predict how soon) where they would be unable to use more carbon fertilizer. Calculations also suggested that more warmth would eventually foster decay in soils, with a net *emission* of greenhouse gases. Around the same time, new evidence showed that the biology of the oceans, too, could vary markedly. The drifting plankton, as complex as a rain forest but scarcely explored by biologists, interacted strongly with the uptake or emission of CO_2.

When people talked now of a GCM they no longer meant a General Circulation Model, built from the traditional equations for weather. GCM now stood for Global Climate Model or even Global Coupled Model, incorporating many things besides the circulation of the atmosphere.

The whole global system was so complex that the old original puzzle—how had ice ages started and ended?—stood unsolved. An important clue came from new ice cores. They testified that as past ice ages had ended, the levels of CO_2 and methane had risen a few centuries *after* the temperature rise. Skeptics seized on this to declare that the greenhouse theory was false, but scientists realized the time lag was not good news. Rather, it confirmed what many had long suspected about the central feedback in the climate system. Evidently the feeble Milankovitch changes in sunlight were enough to tip a balance. Raise the temperature just a bit, and increased emissions from land and sea would pump some greenhouse gases into the air, which would raise the temperature still more . . . and so on. Our modern situation was the reverse: we were initiating changes by raising greenhouse gas levels. That seemed only too likely to set off the feedback process.

Compared with the straightforward physics of how the vast quantities of greenhouse gases that people were adding to the atmosphere blocked radiation, the uncertainties of biology and chemistry in the computer models did not greatly matter. The latest GCMs did seem to be somehow taking all the important factors into account. For they could now reproduce pretty well the Earth's climate under conditions at each step from the middle of an ice age to the present. They got reasonable matches to the observed climate at different levels of solar activity and pollution, including volcanic eruptions. The possibility remained that the models all shared some hidden flaw. That did not mean, as some critics declared, that people need not worry. Broecker and others pointed out that if the models were flawed, the future rise of temperature could be *worse* than predicted.

The biggest problem with the models was that people did not really care much about the average global temperature. Farmers and city managers wanted to know what to expect in their own particular locales. Every model predicted that as the engine of weather drove more heat and moisture around, many regions now dry would get more parched, and many moist places would get more stormy downpours. But the models still could not reliably tell a given locality just what impacts to brace itself for. The IPCC tried to address this problem in 1997 by analyzing "vulnerabilities." If experts could not tell each place whether to expect more rain or less, they could at least say how vulnerable that place would be to either type of change.

To answer that, the experts had to consider not only the region's present climate but also its economic, social, and political conditions. It was not enough, for example, to report that a particular crop would fare worse in warmer weather. For farmers could simply adapt by planting something else, something that grew well under the new conditions. But would they? The various specialists in geophysics now had to talk not only with one another, and not only with experts in other technical fields like agronomy and epidemiology, but also with social scientists, economists, political scientists, and the like.

The IPCC's analysis found, for example, that Africa was "the continent most vulnerable to the impacts of projected changes." That was not just because so many parts of Africa were already subject to droughts and tropical diseases, but still more because population pressure and political failings were bringing environmental degradation that would multiply the problems of climate change. Moreover, Africa's "widespread poverty limits adaptation capabilities."[10]

In Europe and North America, on the other hand, the expertly managed agricultural systems might contrive to benefit from a modest warming and rise in the level of CO_2. To be sure, on every continent nothing could prevent damaging changes in some natural ecosystems. And a sea-level rise would bring costly difficulties along the coasts. There would be threats to health even in the wealthier nations—but they would be able to manage them. An American government study confirmed this relatively mild assessment of the nation's future. In frigid Russia many scientists frankly looked forward to a warmer climate.

These experts could see the consequences as manageable not only by ignoring the world's poor, but also by looking no more than half a century or so ahead. Surely long before then, humanity would have taken control of its emissions, so that CO_2 would not soar to truly dangerous heights . . . wouldn't we? At any rate, the twenty-second century was so far away!

Many climate scientists continued to worry about worse outcomes, less likely but possible. The biggest scientific shock came in 1993 from the remote center of the Greenland ice plateau. Early hopes for a new cooperative program between Americans and Europeans had broken down, and each team set to drilling its own hole. They transmuted competition into cooperation by drilling the two boreholes just far enough apart so that anything that showed up in both cores had to represent a real climate effect, not an accident due to bedrock conditions. The match turned out to be remarkably exact most of the way down. The comparison between cores proved that climate could change more rapidly than almost any scientist had imagined.

Swings of temperature that experts in the 1960s believed must take tens of thousands of years, in the 1970s thousands of years, and in the 1980s hundreds of years, could actually happen in a few decades. During the last glacial period, Greenland had sometimes warmed as much as 7°C in the space of less than fifty years. During the Younger Dryas transition, spectacular shifts in the entire North Atlantic climate were visible within only five snow layers—that is, five years! The evidence could not be dismissed as incredible, for at least one explanation was at hand. Computer models showed that the North Atlantic thermohaline circulation could change drastically, just as Chamberlin had speculated a century earlier (Chapter 1). Meanwhile, geological evidence from other continents suggested that the Younger Dryas had brought climate change not only to the North Atlantic, but all around the planet. (Such events were probably not a global cooling but a "seesaw" in which a change in ocean circulation helped one hemisphere get warmer while the other cooled.)

Could such a climate shock happen not only in glacial times, but also in a warm period like our own? Computer models suggested it could; indeed, we might cause it. As the North Atlantic grew warmer and received more fresh water from rain and melting ice, the thermohaline circulation that moved heat northward could simply stop, changing weather patterns around the planet. Broecker warned that "the ongoing buildup of greenhouse gases might trigger yet another of these ocean reorganizations . . . it could lead to widespread starvation."[11]

Other disturbing news came from the Russians tenaciously drill-

ing ever deeper at Vostok in Antarctica. In their record, now reaching back over four glacial-interglacial cycles, almost every stretch
was peppered with drastic temperature changes. When Bryson,
Schneider, and others had warned that the century or so of stability in recent memory did not reflect "normal" climate variations
(Chapter 5), they had touched on an instability grander than they
guessed.

Despite the profound implications of this new viewpoint, hardly
anyone rose to dispute it. A report written by a National Academy
of Sciences committee in 2001 said that the recognition during the
1990s of the possibility of abrupt global climate change was a fundamental reorientation of thinking, a "paradigm shift for the research community." The IPCC's 1995 report included a notice that
climate "surprises" were possible: "future unexpected, large and
rapid climate system changes (as have occurred in the past)." But
the authors did not emphasize the point. "Geoscientists are just beginning to accept and adapt to the new paradigm of highly variable
climate systems," said the Academy committee in 2001. To everyone
but the climate experts, and even to many of them, future "climate
change" still meant a gradual warming, of limited consequence for
the present generation.[12]

Most politicians barely noticed climate science among the many
urgent demands for their attention. They would not go against
short-term industrial interests unless public opinion drove them.
But debates over global warming, like most political controversies,
did not mobilize the media for long periods. Politicians saw nothing to gain by stirring it up. Even Gore mentioned global warming
only briefly during his 2000 run for the presidency.

Science reporters occasionally found a news hook for a story.
They took mild notice, for example, when the groups compiling
statistics announced that 1995 was the warmest year on record for
the planet as a whole, and when 1997 broke that record, and 1998
yet again. The effect was muted, however, for the warming was

most pronounced in remote ocean and arctic regions. Some smaller but important places—in particular the U.S. East Coast, with its key political and media centers—had not experienced the warming that was evident in many other regions.

Reports of official studies all had their day in the limelight, but rarely more than a day. A story made more of an impression if it dealt with something visible. For example, when tourists who visited the North Pole in August 2000 told reporters that they had found open water instead of ice, news stories claimed that this was the first time the Pole had been ice-free in millions of years. That was dead wrong. Yet the Arctic Ocean ice pack was in fact thinning rapidly; such an incident, by symbolically representing what scientists knew, could be truer than any dry array of data. Other chances to report on climate change came in stories of heat waves, floods, and coastal storms. The better journalists noted that any one of these incidents might have had nothing to do with global warming. Yet the incidents did reflect disturbing statistical trends. And by the late 1990s a few indisputable impacts of global warming had begun to emerge, subtle yet visible to an expert eye. The ranges of some birds and butterflies were shifting northward; the Northern Hemisphere spring was coming on average a week earlier than in the 1970s. To many people, that didn't sound bad.

The world's makers of popular imagery had failed to give the public a convincing picture of global warming. Climate change had featured in a bare handful of science-fiction paperbacks and shoddy movies. Monster storms or a rise of sea level beyond anything actually possible had served only as a background for hackneyed action plots. Prominent skeptics denounced reports of climate research results as just more sensational mythmaking, and many citizens were happy to agree that there was nothing to worry about. The six billion people on the planet had more urgent concerns.

CHAPTER NINE

THE WORK
COMPLETED . . . AND BEGUN

"Some ecological systems, particularly forests . . . may be unable to adapt quickly enough to a rapid increase in temperature . . . most of the nation's coastal marshes and swamps would be inundated by salt water . . . an earlier snowmelt and runoff could disrupt water management systems . . . Diseases borne by insects, including malaria and Rocky Mountain spotted fever, could spread as warmer weather expanded the range of the insects."[1] Those were only some of the threats suggested already in 1988 by a newspaper account of a U.S. government study. Like most reports on the impacts of climate change, it was speckled with terms like "may be" and "could," for it was hard to unravel the tangle of effects, each interacting with others. Arrhenius's first simple inquiry—by how many degrees would average temperature rise?—had expanded into a long list of different kinds of puzzles. They all grew out of one great question: just what would global warming mean for people?

By the late 1990s the world was devoting several billion dollars a year to climate and impact studies. That sounds like a lot, yet it was less than nations spent on many other scientific and technical problems and scarcely matched the scale of the problem. Long gone were the days when the great questions of climate could be profitably studied by a few people taking time off from their usual research. The job now was to find precise answers to countless specific questions. Many of these required costly equipment and teams of highly specialized scientists and technicians.

At the center were the computer modelers, increasingly focusing their work on producing results for the IPCC. As the modelers studied each factor that went into their computations, they exchanged ideas and results with every other specialty that had anything to say about climate. The IPCC reports had become the output of a great engine of interdisciplinary research—a social mechanism altogether novel in its scope, complexity, and significance.

After struggling to a consensus among themselves, the scientists had to reach agreement with government representatives. "It drives you absolutely crazy," one senior scientist complained. "You fly to distant places; you stay up all night negotiating; you listen to hundreds of sometimes silly interventions."[2] In the discussions leading to the IPCC's Third Assessment Report, issued in 2001, new scientific evidence demolished objections from industry-oriented skeptics and persuaded even the most recalcitrant officials. The report bluntly concluded that the world was rapidly getting warmer and declared it likely that this was due mainly to greenhouse gases. Global temperatures would keep on rising at a rate "very likely to be without precedent during at least the last 10,000 years." In some economic scenarios, the late twenty-first century might not see more than an additional degree or two of heat. But if emissions continued to climb unchecked, the average temperature might climb a devastating 5.8°C (10°F).[3]

Among many impacts the IPCC projected, the most prominent was a rise of sea level, up to half a meter by the end of the twenty-first century. As usual, the number came out of a long negotiation that settled on a conservative guess. A minority of scientists warned that the seas could rise as much as a meter. That did not sound like much, but in many areas it would bring the sea inland 100 meters or more (more than 300 feet). Such a rise would not be a world disaster, but by the late twenty-first century coastal areas from Florida to Bangladesh would suffer serious everyday difficulties and occasional calamitous storm surges. Even if emissions somehow stopped at once, the greenhouse gases already in the atmosphere would continue to capture solar energy for a thousand years. Nothing could stop the oceans from continuing to rise, century after century. The last time the planet was 3°C warmer, the sea level had been roughly 5 meters higher, submerging coastlines where hundreds of millions of people now lived.

The IPCC's findings haunted the next mammoth international conference, held at The Hague in late 2000. Although the report was not officially completed, its main conclusions had been leaked to the delegates. Representatives from 170 countries assembled to write the specific rules that might force reductions in greenhouse gases, as promised at Kyoto. Most Europeans demanded a strict regime of greenhouse gas regulation. The U.S. government insisted on more market-friendly mechanisms, and the negotiations collapsed.

When George W. Bush became president, lobbying by his friends in the energy industries persuaded him to throw aside his campaign pledge to restrain greenhouse gases. More, he publicly renounced the Kyoto Protocol. Editorials soundly scolded the action as a surrender to business interests. So it was, but Bush's approach was not far from what a majority of the American public and Congress wanted. To be sure, most people thought it would be good to do something about global warming—but not if that would mean changing anything very much.

Yet even in some corporations, worries about climate change were creeping in. The first powerful group to face the problem head-on was the insurance industry. In the early 1990s many insurers had suffered huge losses as storms and floods increased, much as global warming theorists had predicted. The world's second-largest reinsurance corporation, Swiss Re, warned that companies could be vulnerable to lawsuits if they didn't pay attention to their emissions. A few other European firms, notably the oil giant BP, under the farsighted John Browne, also decided (as he put it in 1997) that "it falls to us to take precautionary action now."[4] Some key American companies acknowledged that greenhouse warming was a real problem and quit the Global Climate Coalition, which collapsed

Such a lobbying organization hardly seemed to be needed, with the interests of the energy business well represented by the Bush administration. Nevertheless, the "Cooler Heads Coalition" (created in 1997) carried on, funded by ExxonMobil and a few other corporations. Wooing legislators and the press, publishing advertisements and offering payments to scientists, the coalition and like-minded conservative groups persuasively denied any need for action. Whether because of these efforts or for other reasons peculiar to the United States, many American policy makers and opinion leaders increasingly diverged from the views held elsewhere. For example, a survey that compared U.S. newspapers with ones in New Zealand and Finland concluded that "the U.S.'s media states that global warming is controversial and theoretical, yet the other two countries portray the story that is commonly found in the international scientific journals."[5]

Scientists were finding a variety of new evidence that something truly exceptional was happening. They were learning to deduce past temperatures not only from historical records of events like freezes and harvests, but also from analysis of isotopes in tree rings, coral reefs, deposits in caves, and so forth. An example of these far-flung efforts was a series of perilous expeditions that labored high in the thin air of the Andes and Tibet to drill into tropical ice caps. (During one storm the leader, Lonnie Thompson, saved his tent from blowing off the mountain by stabbing his ice-axe into the snow through the tent floor.) The new ice cores showed that global warming in the last few decades had been greater than anything seen for thousands of years. Indeed, the tropical ice caps themselves were melting away faster than scientists could measure them. A graph that compiled estimated temperatures over the past ten centuries, featured in the 2001 IPCC report and widely reprinted, showed a sharp turn upward since the start of the industrial revolution. It seemed that 1998 had been not just the warmest year of the century, but of the millennium.

People who wanted to raise awareness of global warming seized on the graph, to the regret of some climate experts who recognized that, like all new research, the results were uncertain. There is nothing simple about any set of data, for many steps of analysis are needed between the first measurement and the final number. For example, it may seem straightforward to read a thermometer every day at noon, but what if someone forgets to make an adjustment when daylight savings time begins? Thousands of data sets in the history of climatology had been polluted by such errors, trivial or subtle. Worse, the popular press often displayed the historical temperature graph as a single, stout, level line that hooked up at the end like a hockey stick. The original graph had a broad gray band of shading to show a range of uncertainty, which might conceal big climate shifts (Figure 3).

Global warming deniers had little choice but to attack the "hockey stick" as an error, or even a deliberate fraud. Some argued that there had been a Medieval Warm Period just as hot as the twentieth century. Expert climatologists looking into the period, however, found a scattering of warm and cold spells in different places at different times. Like the temporary cooling of the 1960s, the Medieval Warm Period took place mainly in some parts of the Northern Hemisphere, unlike the recent warming over most of the planet. Several groups now worked out their own reconstructions of past temperatures; the best way to determine whether a set of data is mistaken is to dig up other data. In scientific publications the single "hockey stick" line gave way to sets of lines like a tangle of

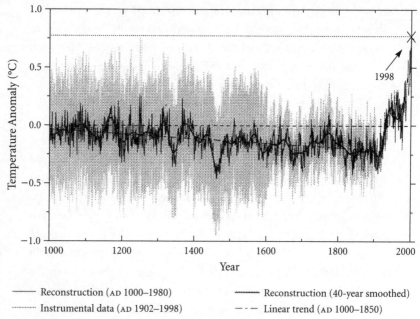

Figure 3. THE UNPRECEDENTED RECENT RISE OF GLOBAL
TEMPERATURE.

The "hockey stick" graph: a reconstruction of Northern Hemisphere temperatures for the past millennium, and measured temperatures for the past century.
The gray band shows the spread of the data; the dark line, an average, gave a misleading impression of certainty. More recent reconstructions wander up and
down through the gray area, but all agree that the rise since the mid-twentieth
century is unprecedented. (M. Mann et al., *Geophysical Research Letters* 26
[1999]: 761, copyright 1999 American Geophysical Union, reproduced by permission.)

spaghetti. Every line hooked up sharply in the last half century. The
predictions that scientists like Hansen had been making since the
1970s were turning out to be right.

The diminishing cluster of scientists who clung to the belief
that the greenhouse theory was wrong continued to come up with

fragments of evidence that seemed to confirm their position. One of their arguments was strong enough to draw the attention of other scientists. Computer models predicted a greenhouse warming "fingerprint" for how temperature would increase at each level of the atmosphere. A data set from a satellite contradicted this, showing no warming at middle levels, perhaps even a cooling. A set of balloon measurements also showed no warming at those levels. The computer modelers, try though they might, could not force their GCMs to reproduce the effect. Yet few experts thought this refuted the entire idea of global warming. They knew that it took a highly complex analysis to convert the numbers radioed back by the satellites or the balloons into actual air temperatures. The new data sets could lower experts' belief in the greenhouse theory by only a small percentage when so many other lines of evidence pointed firmly in the opposite direction.

Especially persuasive was a historical survey that dug into thousands of old ships' logs for records of ocean temperature. The seas absorb many times more heat energy than any other component of the climate system; it was here if anywhere that a warming should be visible. The heat content of the upper levels had indeed risen markedly in the second half of the twentieth century. And the temperature patterns at various levels in different ocean basins neatly matched the "fingerprints" that ocean-atmosphere GCMs calculated for the greenhouse effect.[6]

Meanwhile, intensive study of the satellite and balloon records turned up hidden mistakes in the analysis of each. The middle levels of the atmosphere had in fact been warming up, more or less as greenhouse-effect models had predicted.

For the public, more striking than any graph were then-and-now photographs that showed glaciers receding everywhere from the Alps to the fabled snows of Kilimanjaro. Deniers fastened on one or another case that offered contrary evidence, but in fact more than nine-tenths of the world's mountain glaciers were shrinking. Ice was melting back to reveal mummies that had been frozen for thousands of years.

Real events were shoving into the abstract studies of potential impacts. It turned out that because of unexpected complexities, the rich nations were not as safe as many had thought. For example, in 2003 a heat wave surpassing anything in the historical record struck Europe. Calculations showed the heat was probably intensified by global warming. Nobody had foreseen that old people could not save themselves when the traditional August vacation emptied the cities; tens of thousands died. Another example: bark beetles, no longer controlled by winter freezes, began to devastate millions of acres of forests from Alaska to Arizona. The weakened timber fell prey to an unprecedented outbreak of forest fires. Nobody had prepared for this particular consequence of global warming.

Public understanding, always shaky on scientific topics, kept up fairly well with the emerging consensus of experts. Since the surge of media attention in 1988, polls had found that roughly half of Americans thought global warming was already here, and many of the rest thought it was coming. But most people in all countries surveyed admitted that they knew little about climate change, and few held strongly to their views. The fraction of Americans who claimed they worried about global warming "a great deal" actually declined in the early 2000s, and by 2004 a bare majority in the United State expressed any worry at all.[7]

Western Europeans meanwhile grew *more* worried. They had been gradually pulling ahead of Americans in all environmental concerns, and even conservative European political parties increasingly agreed to regulate various kinds of pollution. After the deadly heat wave of 2003, some governments began to discuss serious restrictions.

Then came the summer of 2005, the worst Atlantic hurricane season on record, capped by Hurricane Katrina's devastation of New Orleans. "Are We Making Hurricanes Worse?" asked the cover of *Time* magazine on 3 October. Scientists were in fact divided on that, and they furiously debated whether global warming had increased the hurricane risk. It was another case where an event that was not necessarily caused by global warming nevertheless taught an accurate lesson, for rising sea levels would eventually lift storm surges over existing levees. But what really struck the public was images. The half-submerged buildings of science fiction, the "environmental refugees" that experts had been foreboding for decades, filled television screens in real time.

International polling now found that almost everywhere in the world, a majority of the population had heard of global warming. Of these, in most nations a quarter to a half felt "a great deal" of concern, substantially more than just a few years earlier. Even in China many were coming to believe that climate change endangered their economic progress and social stability.

Polls and focus groups, however, found that most people scarcely connected the greenhouse effect with their daily lives. Asked to name environmental problems facing the nation, most Americans would think of pollution of drinking water, local smog, or the destruction of tropical forests ahead of climate change. Europeans usually ranked climate change first among environmental worries, but well below their worries on various other issues. Many who expressed concern thought nothing they could do would make a difference anyway, and others were satisfied with small, symbolic steps. Many did not even understand that fossil fuels were the chief source of global warming. Some hoped that futuristic technical advances would somehow fix any problems. One study of Americans concluded that most of them, anxious and baffled, literally did not like to think about the subject at all. "Their concern translates into frustration rather than support for action."[8]

In many countries smaller fractions of the public, perhaps 5–10 percent, held stronger views. On one side stood people who feared a grave, even apocalyptic threat to their way of life and perhaps all life on the planet. On the other side stood people who dismissed global warming as a myth concocted by self-serving intellectuals. If you guessed that a member of the first group leaned politically to the left, and a member of the second group to the right, you would be correct three-quarters of the time—at least in the United States, where the issue was more politically polarized than elsewhere.

Both extremes found comfort in sympathetic media productions. One example was *State of Fear,* a science-fiction thriller that became a best seller in 2005. The author, Michael Crichton, portrayed the scientific establishment as arrogant and untrustworthy, quick to make any claim that would bring research funds. Like-minded Web sites and blogs scorned climate warnings as leftist propaganda and passed around tidbits of contrary research findings. There are always anomalies at the research front, and as soon as scientists resolved one doubt, the deniers fastened on a newer one. Better-informed conservatives focused on policy. Nothing we could do to mitigate global warming, they said, would make a difference worth the enormous costs.

On the other side, some people frankly tried to spread their fears about climate through their own Web sites, blogs, and media productions. Forebodings became vivid in "The Day after Tomorrow," a 2004 special-effects movie. Perhaps a tenth of the American public and many elsewhere saw this tale of impossible cataclysms, including an ice age that struck in the course of a few days when the Gulf Stream halted. That sounded like the way popular science articles and television specials had described the shutdown of the North Atlantic thermohaline circulation, what Broecker and others had been speculating about since the 1980s. After all, England is as far north as Labrador, so without heat from the Gulf Stream, wouldn't it get as cold?

A few scientists belatedly took a closer look and announced that nothing so dreadful was possible. If the thermohaline circulation (a much broader movement of waters than the Gulf Stream) ever did cease, that would bring no ice age. Labrador is downwind from tundra that freezes each winter, whereas England is kept warm by prevailing westerly winds that pick up the ocean's heat, most of it simply retained from the summer. Improved computer models calculated that although the circulation was indeed likely to slow down during the next century or two, the slowdown should be gradual. Any cooling that it brought would probably be outweighed by greenhouse warming. Broecker himself criticized the "exaggerated scenarios" in the media.[9]

Professional science journalists struggled to find accurate but interesting ways to explain the latest scientific reports. For decades television had illustrated climate change with stock footage of drought-struck crops, hurricane blasts, or advancing waves; political cartoonists sketched half-submerged buildings or whirling tornadoes. Familiarity blunted such images, long associated with ordinary weather problems. New pictures showed up in the early 2000s: glaciers dropped icebergs into the sea, Pacific islanders and Alaskan seal hunters complained about threats to their ways of life, polar bears struggled to survive as ice floes melted. All that lay far from the lives of most citizens. Many in the industrialized nations saw global warming as a remote abstraction, not so much a problem for folks like themselves as for tribal natives and polar bears.

Nobody had produced a significant novel or movie that humanized the travails that climate change was realistically likely to bring upon us—the squalid ruin of the world's mountain meadows, coastal wetlands, and coral reefs, the spread of tropical diseases to new locales, the press of millions of starving refugees from drought and seacoast flooding, perhaps entire nations driven to aggression or despair.

What did draw universal attention was a shoestring-budget documentary film. Since 1990 Al Gore had been developing an illustrated lecture on the global warming story. Converted into a film titled *An Inconvenient Truth,* in the year following its 2006 opening the presentation became famous around the world. It made a strong impression on the sort of people who saw documentaries, including many key policy makers.

Larger forces were pushing in the same direction. As the meaning of the 2001 IPCC report sank in, American science reporters and their editors began to recognize where the scientific debate really stood. Some admitted that they had leaned over too far in granting "equal time" to the small remnant of deniers. Coverage of climate change in American newspapers, which had declined in the mid-1990s to well below the European level, began to climb back. After Hurricane Katrina, respected media like *National Geographic* magazine and the Weather Channel risked angering a segment of their audiences by featuring global warming as a truly serious problem. "Be worried," the cover of *Time* advised. "Be *very* worried." In 2006 a reporter mused that "global warming has the feel of breaking news these days."[10]

Public opinion responded only sluggishly, but important leadership groups shifted. Rock musicians and movie idols worked seriously to warn about climate change. Devout Christian climate scientists persuaded some key evangelical leaders that we were called upon to protect God's creation from greenhouse warming. The chief executives of corporate heavyweights like General Electric and Wal-Mart set out with the vigor of converts to limit their companies' emissions.

Business Week called 2006 "the year global warming went from controversial to conventional for much of the corporate world."[11] Executives who held back felt pressure from several directions. A pledge to fight climate change would improve their corporate image. Some corporations faced actual lawsuits for the damage their greenhouse emissions could cause. Wall Street investors began to weigh global warming risks when evaluating a company. Most important, legal restrictions on emissions seemed inevitable. As the *Wall Street Journal* reported, "The biggest question going forward no longer is whether fossil-fuel emissions should be curbed. It is who will foot the bill for the cleanup." A savvy corporation would take the lead in discussing how to tax or regulate greenhouse gases. If you're not at the table, the *Journal* remarked, you're on the menu.[12]

Although governments had never given much money for studies of energy conservation or renewable energy sources, experts were turning up ways, practical and even profitable, to start restraining climate change. For example, fixing the leakage of methane gas from pipelines would save money while removing one source of global warming. Cutting back the emission of microscopic soot particles from coal-burning plants would not only reduce the warming that the black particles caused, but reduce the expense of the widespread health problems they also caused. Overall, greater efficiency and less pollution would not weaken the economy, but strengthen it.

Even the Bush administration showed signs of a policy change. In his 2007 State of the Union Address, the president finally agreed that we should "confront the serious challenge of global climate change." But the voluntary actions he proposed would not make a great difference. Other levels of society acted more effectively. Thoughtful officials in fields ranging from municipal water supplies to forestry began to plan for a changed world. By 2007 Democrats in dozens of states were joining with Republicans like the governors of California, Florida, and New York to plan mandatory systems to cut greenhouse emissions. Hundreds of big cities around the world were studying how they should alter everything from automobile traffic to building codes. Churches, universities, and countless other nonprofit organizations joined in, along with a rapidly increasing number of individual citizens. Many pledged to try to meet the Kyoto Protocol target for reducing emissions.

The negotiators in Kyoto had agreed that the Protocol would not go into effect until it was ratified by nations causing more than 55 percent of the world's CO_2 emissions. With the United States government standing aside, that final step was not taken until Russia joined in 2005. Meanwhile, at an international meeting held in 2001 in Bonn, 178 governments—but not the United States—had worked out measures to implement the Protocol. The goal, watered down from the original one, was to return greenhouse gas emissions to their 1990 rate by 2010.

Scarcely anyone believed the goal could be achieved, or that achieving it would do much to slow global warming. Diplomacy is a gradual process. The first and most important work is to shift attitudes step by step. That would clear a path for devising legal mechanisms such as procedures to measure national emissions and to adjudicate quotas. These would lead eventually to changes in actual facts. From the start, the aim was to motivate people to work out and adopt better procedures and technologies. Already projects like wind farms in Denmark and solar-powered generators in Spain were showing promise. On the legal front, in 2005 the Europeans adopted a scheme of mandatory permits for carbon emissions. The system was so badly implemented that the price of the permits soared, then abruptly crashed to almost nil. Meanwhile, "offsets"—whereby, for example, a power utility in Britain was allowed to emit CO_2 if it paid a utility in China to reduce its own emissions—suffered from widespread errors and blatant fraud. In a perverse way all this was what the Kyoto delegates had wanted: experiments to find how a particular policy worked in practice. Such experiences could lay a foundation for negotiating the new treaty that would need to be in place when the Kyoto Protocol expired in 2012.

That round of negotiations waited on the next IPCC report. Computer modelers, struggling with their huge responsibility, remained at the convergence of many research programs. There were now about a dozen major teams running formidable ocean-atmosphere GCMs on supercomputers. Intractable problems of cloud and aerosol behavior still kept them from making reliable regional projections, but the modelers' global figures were converging. And they now had an entirely independent check. Geologists had ingeniously deduced levels of both temperature and CO_2 clear back to the time of the dinosaurs, finding large fluctuations from era to era. By 2006 the geologists had a rough number for the correlation: a 3°C rise whenever CO_2 had doubled. The match with the GCMs' number was a welcome confirmation for scientists, but not good news for human society.

Some research did bring good news. The headlong surge of methane into the atmosphere seen in the 1980s had come to a halt (experts disagreed on why, and whether it was a halt or only a pause). Also, rough calculations dispelled some of the worry about methane released from clathrate ices in the deep oceans. To be sure, geologists had turned up evidence that huge clathrate eruptions could have been involved in a catastrophic climate change and mass extinction 55 million years ago. But for clathrates as for the North Atlantic thermohaline circulation, the latest computer ocean models suggested nothing terrible was likely to happen in the next century or two—and of course anything beyond that was too far away to be worth speculating about.

Most of the scientific news, however, was bad. The world's emissions of CO_2 and some trace greenhouse gases were accelerating at a rate that matched the IPCC's most pessimistic scenarios. That was thanks to the stunning industrial growth of China and other developing nations, along with the failure of most developed nations to fulfill their Kyoto pledges. The planet's natural geophysical and biological systems were not likely to help. In 2000 researchers had managed to couple together computer models for the atmosphere, oceans, vegetation, and soils. They predicted that around mid-century the planet's warmer biosphere would probably turn from a net absorber to a major *emitter* of carbon, which would speed up climate change.

Within a few years there were reports that feedbacks long anticipated were already kicking in. A decrease of Arctic snow cover and sea ice was making for more absorption of summer sunlight. The oceans, too, were getting less efficient at taking up CO_2 as the water grew warmer and more acidic. Some forests that used to absorb carbon were drying out and burning instead. Possibly the worst threat came from tundra, where researchers saw greenhouse gases starting to bubble out of bogs before their eyes. A Siberian researcher called it an "ecological landslide that is probably irreversible and is undoubtedly connected to climatic warming."[13] Around 2005 the phrase "tipping point" began to show up frequently in scientific papers and media reports on climate, an admission that it might soon become impossible to arrest such changes.

Most startling of all was the dwindling of the Arctic ice pack, just as computer models had predicted—only faster. Within a few decades the Arctic Ocean might be entirely free of ice in summer, for the first time in millions of years. Effects of warming appeared even in Antarctica, which scientists had thought would be protected for centuries by its encircling ocean and colossal ice cap. Some parts were indeed not warming or even sporadically cooling, but other parts were losing ice far in advance of expectations. Entire floating ice shelves that had been in place for thousands of years were disintegrating. Adventurous surveys across the ice confirmed the old speculation that removing an ice shelf could dramatically accelerate the drainage of glaciers corked up behind it. Theorists struggled to fit computer models to the new data. "The response time scale of ice dynamics is a lot shorter than we used to think it was," admitted a leader of the research.[14] Scientists were less confident than ever about how many centuries it might take to drain the entire West Antarctic Ice Sheet.

New data hinted that the Greenland Ice Sheet, too, was less stable than most experts had thought. As summers grew warmer, the snow on the surface was getting wet, becoming darker and absorbing more sunlight. Like Antarctica but faster, Greenland was losing ice around its edges sooner than expected. Conceivably, surging ice streams might raise the sea level at a catastrophic rate sometime within the next century or two. Unless we act promptly, declared Hansen, we might find ourselves living on a "different planet."

In 2007 the IPCC issued its Fourth Assessment Report. The computer modelers had grown more certain that we were unlikely to get away with a rise of less than 1.5°C by the last decade of the century. The models could not agree so well on an upper limit, leaving open a small but all too real possibility that global temperature could soar to a disastrous 6°C or even more. The complexities of climate stubbornly refused to allow precision, for nobody could accurately measure the many feedbacks until the world actually warmed up.

The scientists did feel more certain about some things. First, that serious effects of global warming were now upon us. Around the world they were seeing worse heat waves, droughts, and stormy rains. Some places were enjoying their new climate, and some crops were benefiting from CO_2 fertilization and warmer nights. But pests and tropical diseases were spreading to warmed regions. The oceans were becoming more acidic, at a rate that would probably harm fisheries and eventually annihilate coral reefs.

In the exhausting sessions where government representatives argued out the IPCC's crucial "Summary Report for Policymakers," the delegates agreed that human responsibility for the warming was "very likely," meaning between 90% and 99% certain. Most of the scientists would have called it at least 99% certain. Other conclusions, too, were toned down until even the most cautious could accept them. For example, the panel on sea level again projected a rise of no more than half a meter by 2100. Only an inconspicuous note explained that this was calculated from normal melting and thermal expansion, taking no account of the untested ideas on how ice sheets might surge rapidly into the oceans.[15]

The sea level had actually been climbing at the very upper limit of the range that IPCC reports back to 1990 had projected. The same was true of the actual temperature rise. One might say the IPCC process had been soberly conservative when it negotiated projections that ranged from negligible up to the greatest likely changes. Some, however, thought that being conservative meant attending to risks that were less likely but potentially catastrophic—as people do, for example, when they budget for armed forces or build a nuclear reactor. As one geophysicist told his colleagues, "Up until now many scientists may have consciously or unconsciously downplayed the more extreme possibilities at the high end of the uncertainty range, in an attempt to appear moderate and 'responsible' (that is, to avoid scaring people). However, true responsibility is to provide evidence of what must be avoided."[16]

Such arguments weighted on the IPCC authors as they wrote a final synthesis of the panel's 2007 reports, issued late in the year. This would be the end point of their work and the starting point for the next long round of international negotiations. The panel was now better known and better respected for having shared a Nobel Peace Prize with Al Gore, and the authors ventured to put a spotlight on the risk estimates that had been buried in the long, technical reports. With CO_2 and other greenhouse gas emissions rising at an accelerating rate, climate change was likely, for example, to put a quarter of the world's species at risk of extinction. Still more likely would be, for example, "disruption of . . . societies" by storm floods. Less certain but perhaps more important was the possibility of "abrupt or irreversible" effects. In particular, damaging "sea level rise on century time scales cannot be excluded." If greenhouse gas levels kept rising unrestrained, well beyond twice the preindustrial level, we would probably see a radical impoverishment of many of the ecosystems that sustain our civilization.[17]

This was no inevitable fate, but a challenge. Economists were finding that we could bring global warming to a halt with existing or easily developed technologies. It was not hard to see where to begin. The world's governments were still *subsidizing* global warming, devoting far more money to supporting the use of fossil fuels than to developing alternative sources of energy. The most influential study came in 2006, worked up for the British government by a team of experts headed by Nicholas Stern, a former chief economist of the World Bank. They estimated that if the warming was in the upper range of what scientists thought likely, by the end of the century it could cut the annual Global Domestic Product (GDP) by 5 percent and perhaps even more. The damage could eventually be comparable to the devastation of the Second World War. The cost of preventing this would most likely be a modest 1 percent reduction in GDP, equivalent to a delay of a few months of growth. "Climate change," Stern said, "is the greatest market failure the world has ever seen."[18]

Some found a still more sobering way to frame climate change: as a security threat. Military officers saw that the armed forces might be called on to handle anything from hordes of environmental refugees to outright combat over dwindling resources. Senior officers began to say that among long-term dangers, global warming ranked at least as high as terrorism. Some political leaders agreed, along with a majority of the world's public.

This was the discovery of global warming by ordinary people and, more important, by leadership elites. The scientific discovery itself had been completed in the years leading up to the IPCC's 2007 report. Now many scientists, worn out from giving as much as half their time for years to the international process on a voluntary basis, wanted to turn to the details of climate change and how to address it. The IPCC had fulfilled the historic duty for which it was created. It had shown that the science of global warming *mattered*, in everyday human terms. From technical discussions about X°C the world had moved on to agenda items for action.

Predictions for global temperature toward the end of the century still stood at a damaging 3°C rise—more or less. It was impossible to rule out a fairly benign rise; it was impossible to rule out a wholly catastrophic rise. Yet the central question had found its answer, an evaluation of danger far more precise than what people will act on when faced with an economic crisis or a foreign threat. The world had completed an extraordinary journey. The hypothesis proposed by Arrhenius in 1896, rejected by almost every expert through the first half of the twentieth century, advancing steadily through the second half, was now as well accepted by scientists, policy makers, and the public as any statement of future risks ever could be.

The biggest source of uncertainty no longer lay in the science. To predict climate change, you would first have to predict changes in CO_2, methane, and other greenhouse gases, as well as emissions of smoke and other aerosols, not to mention changes in forests and other ecosystems. These changes now depend less on geochemistry and biology than on human decisions. Whether the world will see a mild or a drastic climate change depends chiefly on future social and economic trends, and above all on control of emissions. In the IPCC reports, scientists gave their best answer. Now the main question is what people will choose to do.

When should we believe that scientists are giving us reliable information about the world? Most stories of the "advance" of science evoke a picture of people marching resolutely ahead. A scientist "discovers" something, like an explorer of old who first comes into an unknown valley. Other explorers push onward, each taking knowledge a step forward. That would be "progress" in the old meaning of the word, a stately parade advancing according to plan. But in reality, after a scientist publishes a paper with a new idea or observation, other scientists usually look on it with justifiable suspicion. Many papers, perhaps most of them, harbor misconceptions or plain errors. After all, research (by definition) operates past the edge of the known. People are peering through fog at a faint shape, never seen before. Every sighting must be checked and confirmed.

Scientists themselves find the most convincing confirmation of an idea is one that comes from the side, from some other team using an entirely different type of observation or line of thought. Such connections among different realms are especially common in a science like geophysics, whose subject is intrinsically complex. Scientists may start with an idea about the smoke from volcanoes, put it alongside telescopic observations of Venus, notice the chemistry of smog in Los Angeles, and plug it all into a computer calculation about clouds. You cannot point to a single observation or model that convinced everyone about anything.

This doesn't look like an expedition moving into new territory. It looks more like a crowd of people scurrying about, some huddling together to exchange notes, others straining to hear a distant voice or shouting criticism across the hubbub. Everyone is moving in different directions, and it takes a while to see the overall trend. I believe this is the way things commonly proceed, not only in geophysics but in most fields of science.

In this book I have tried to show this process by connecting the dots among more than a thousand of the most important papers in the science of climate change.[1] For each one of these select thousand, scientists published another ten or so papers of nearly the same importance, describing related data, calculations, or techniques. And for each of those ten thousand, specialists in that particular subject had to scan at least ten other publications that turned out to be less significant—studies that offered minor corroborations, or perhaps contained distracting errors, or turned out not to be relevant at all. By pulling the main developments above the tumult, this book gives a clearer picture than scientists could see at the time.

Getting coherent explanations is harder in geophysics than in such relatively self-contained disciplines as astrophysics and molecular genetics. Scientists in those disciplines address problems that fall within a well-understood boundary. That boundary is roughly congruent with a social boundary, defining a community with its own journals, scientific societies, meetings, and university departments. Scientists develop these social mechanisms partly to facilitate their work of training students and raising funds for research. Still more, the social coherence of the discipline is invaluable in their work of communicating findings to one another, debating them, and reaching conclusions about which findings are reliable.

For the process to work, scientists must trust their colleagues. How is this trust maintained? Telling the truth is important, but it is not enough: although scientists rarely cheat one another, they easily fool themselves. The essential kind of trust comes from sharing a goal, namely, the discovery of reliable knowledge, and from sharing principles about how to pursue that goal. One necessary principle is tolerance of dissent, encouraging every rational argument to be heard in public discussion. A second principle is limited consensus, working to agreement on important points, even while disagreeing on others.

Maintaining trust is more difficult where the social structure is fragmented. A community in one specialty cannot thoroughly check the work of researchers in another, but must find a few people who seem reliable and accept their word for how to view the latest results. The study of climate change is an extreme example. Researchers cannot separate meteorology from solar physics, pollution studies from computer science, oceanography from glacier-ice chemistry, and so forth. The range of journals they cite in their footnotes is remarkably broad. So many different factors influence climate! But this complexity imposes difficulties on those who try to reach solid conclusions about climate change. Establishing a level of reliability depends on a process of checking and correction by dozens of scientific communities, each dealing with its own piece of the problem.

Who made the discovery of global warming—that is to say, the discovery that human activities are making the world warmer? No one person, but a sprawl of scientific communities. Their achievement was not just to accumulate data and perform calculations, but also to link these together. This process was so complex and so important that its late stages were visibly institutionalized: the workshops, reviews, and negotiating sessions of the Intergovernmental Panel on Climate Change. The discovery of global warming was patently a social product, a limited consensus of judgments arising in countless discussions among thousands of experts.

When people hear of a discovery, they make an implicit assessment of its reliability, that is, how strongly they should believe it is true. The IPCC was pressed to be explicit. When its members announced in 2007 that they found it "very likely" that the current unprecedented rate of warming was largely due to our emission of greenhouse gases, they explained that this meant they judged it was between 90 and 99 percent probable that this was true.[2]

Some people continued to believe that "global warming" was nothing but a social construction—more like a myth invented by a community than a fact like a rock you could hold in your hand. After all, the critics pointed out, communities of scientists had often held mistaken views and then changed their collective minds. Most scientists found this denial not persuasive—and not even interesting. To be sure, half a century back, most scientists had found Callendar's greenhouse-warming proposition implausible. But scientists back then had understood that their ideas about climate change drew on a mere scattering of hand-waving arguments and uncertain measurements. Although Callendar's proposition flew in the face of ideas about climate stability that scientists had long taken for granted, experts set it aside only provisionally. It stuck in their minds, awaiting the coming of better data and theories. The views of the public and of the scientific community changed together, each acting on the other. On the public side, hard experience drove home how severely our technologies can change everything, even the air itself. Meanwhile, on the scientific side, knowledge of how climate could change evolved under the influence of countless field observations, laboratory measurements, and numerical calculations, yet within limits set by the entire society's commonsense understanding (and by the funding it provided). Eventually the conclusions were solid enough to be perpetuated in consensus panel reports. By the end of the twentieth century it was the critics who clung to traditional beliefs, the fantasy of a benignly normal and self-regulating climate.

In a restricted sense, one could call the present understanding of climate change a product of human society. We should not call it *nothing but* a social product. Future climate change in this regard is

like electrons, galaxies, and many other things not immediately ac-
cessible to our senses. All these concepts emerged from a vigorous
struggle of ideas, until most people were persuaded to say the concepts represented something real. Just what "real" meant was open
to debate; philosophers offer many opinions about how a scientific
concept may correspond to an ultimate reality. That ageless question rarely troubled climate scientists, who took it for granted that
the future climate is as real as a rock. At the same time, scientists
readily admitted that their knowledge of this future thing could be
stated only within a range of probabilities.

Our understanding of climate goes beyond scientific reports into
a wider realm of thinking. When I look at the maples on my street,
still green in late November for the first time in memory, I may see
a natural weather variation, or I may see a human artifact caused by
greenhouse gas emissions. Such perceptions are shaped not only by
scientists, but also by interest groups, politicians, and the media.
For global warming in particular, the social influences run deeper
still. Unlike, say, the orbits of planets, the climate in the future actually does depend in part on what we think about it. For what we
think will determine what we will do.

What can we do about global warming, and what should we do?

My training as a physicist and historian of science has given me
some feeling for where scientific claims are reliable and where they
are shaky. Of course, climate science is full of uncertainties, and nobody claims to know exactly what the climate will do. That very uncertainty is part of what, I am confident, is known beyond doubt:
our planet's climate can change, tremendously and unpredictably.
Beyond that we can conclude (with the IPCC) that it is *very likely*
that serious global warming, caused by human actions, is coming in
our own lifetimes. The probability of harm, widespread and grave,
is far higher than the probability of many other dangers that people

normally prepare for. The few who contest these facts are either ignorant or so committed to their viewpoint that they will seize on any excuse to deny the risk.

Thanks to the strenuous labors by thousands of people described in these pages, we have had a warning in time—although just barely in time, for governments have delayed in heeding the warning. If there is even a small risk that your house will burn down, you will take care to install smoke alarms and buy insurance. We can scarcely do less for the well-being of our society and the planet's ecosystems. Thus, the only useful discussion is about what measures are worth undertaking.

Many things can be done right now that are not only cheap and effective, but will actually pay for themselves through benefits entirely unrelated to their action against global warming. A good start would be to remove all government subsidies for fossil fuels, which are staggeringly large, mostly hidden, and economically unsound. For the United States, another sensible step would be to gradually raise the tax on gasoline by a few dollars (comparable to what nearly all other industrial nations pay, and compensated for by lowering other taxes) to cover the actual costs of roads, traffic congestion, and medical care for accident injuries and illness due to smog. Other economically beneficial policies could improve fuel efficiency in many areas, protect forests, and so forth. Looking beyond CO_2, money can actually be saved while reducing the greenhouse effect by attacking unhealthy smoke emissions and other environmental harms. Such steps can be taken by local as well as national governments, and by most businesses and individual citizens. Americans in particular—responsible for far more of the greenhouse gases now in the atmosphere than any other group, and the people best placed to do something about it—must set an example.

Most important of all, taxes and regulation will create sensible "price signals" that will stimulate development of technologies and practices that can advance national economies with far lower greenhouse gas emission. A good bit of that development is already under way, but technologies do not magically grow by themselves. According to economic demands, a technology may remain stagnant or dash forward to solve problems with remarkable speed. The control of CFCs, for example, turned out to be far easier and cheaper than the regulated industries feared.

To say that such steps are socially or politically impossible is to forget that far greater changes have come swiftly, in countless areas, once people set their minds to change. (Think how our patterns of living, even of eating, have altered over the past fifty years!) Citizens can reconsider their personal practices and put pressure on businesses and governments. This is not a job for someone else, sometime down the road: we have already run out of time. The first practical steps, the easiest ones, will not have a great influence on future global warming. But every step will bring experience in developing and negotiating effective technologies and policies. We will need this experience when, as is likely, increasingly grave harm from climate change drives us to greater efforts. If we act now, the job will be far cheaper and easier than it was to win the Second World War or the Cold War, and more likely to inspire the best in us, too.

Like many threats, global warming calls for increased government activity, and that rightly worries people. But in the twenty-first century, in many areas the alternative to government action is not individual liberty; it is corporate power. And the role of large corporations in this story has been mostly negative, a tale of self-interested obfuscation and shortsighted delay. The atmosphere is a classic case of the problem of the "commons": in the old shared English meadows, any given individual was bound to gain by adding more of his own cows, although everyone lost from the overgrazing. In such cases the public interest can be protected only by public rules.

It is nearly certain that global warming is upon us. It is prudent to expect that weather patterns will continue to change and the seas will continue to rise, in an ever-worsening pattern, through our lifetimes and beyond. Nearly everyone in the world will need to adjust. It will be hardest for the poorer groups and nations among us, but nobody will be exempt. Citizens will need reliable information, the flexibility to change their personal lives, and efficient and appropriate interaction at all levels of government. So it is an important job, in some ways our top priority, to improve the communication of knowledge and to strengthen democratic control in governance everywhere. The spirit of fact gathering, rational discussion, toleration of dissent, and negotiation of an evolving consensus, which has characterized the climate science community, can serve well as a model.

1800–1870 First Industrial Revolution. Coal, railroads, and land clearing speed up greenhouse gas emission, while better agriculture and sanitation speed up population growth.

The level of carbon dioxide gas (CO_2) in the atmosphere, as later measured in ancient ice, is about 290 ppm (parts per million). The mean global temperature (1850–1870) is about 13.6°C (56°F).

1824 Fourier calculates that the Earth would be far colder if it lacked an atmosphere.

1859 Tyndall discovers that some gases block infrared radiation. He suggests that changes in the concentration of the gases could bring climate change.

1896 Arrhenius publishes the first calculation of global warming from human emissions of CO_2.

1897 Chamberlin produces a model for global carbon exchange that includes feedbacks.

1870–1910 Second Industrial Revolution. Fertilizers and other chemicals, electricity, and public health further accelerate growth.

1914–1918 World War I; governments learn to mobilize and control industrial societies.

1920–1925 The opening of the Texas and Persian Gulf oil fields inaugurates an era of cheap energy.

1930s A global warming trend since the late nineteenth century is reported.

Milankovitch proposes orbital changes as the cause of ice ages.

1938   Callendar argues that CO_2 greenhouse global warming is under way, reviving interest in the question.

1939–1945   World War II; the grand strategy is driven largely by a struggle to control oil fields.

1945   The U.S. Office of Naval Research begins generous funding of many fields of science, some of which happen to be useful for understanding climate change.

1956   Ewing and Donn offer a feedback model for quick ice-age onset.

              Phillips produces a somewhat realistic computer model of the global atmosphere.

              Plass calculates that adding CO_2 to the atmosphere will have a significant effect on the radiation balance.

1957   The launch of the Soviet *Sputnik* satellite. Cold War concerns support the 1957–1958 International Geophysical Year, which brings new funding and coordination to climate studies.

              Revelle finds that CO_2 produced by humans is not readily absorbed by the oceans.

1958   Telescope studies show a greenhouse effect raises the temperature of the atmosphere of Venus far above the boiling point of water.

1960   Mitchell reports a downturn of global temperatures since the early 1940s.

              Keeling accurately measures CO_2 in the Earth's atmosphere and detects an annual rise. The level is 315 ppm. The mean global temperature (a five-year average) is 13.9°C.

1962   The Cuban Missile Crisis, the peak of the Cold War.

1963   Calculations suggest that feedback with water vapor could make the climate acutely sensitive to changes in the CO_2 level.

1965   At a Boulder, Colo., meeting on the causes of climate change, Lorenz and others point out the chaotic nature of the climate system and the possibility of sudden shifts.

1966   Emiliani's analysis of deep-sea cores shows the timing of ice ages was set by small orbital shifts, suggesting that the climate system is sensitive to small changes.

1967   The International Global Atmospheric Research Program is established, mainly to gather data for better short-range weather prediction, but climate research is included.

            Manabe and Wetherald make a convincing calculation that doubling CO_2 would raise world temperatures a couple of degrees.

1968   Studies suggest a possibility of collapse of Antarctic ice sheets, which would raise sea levels catastrophically.

1969   Astronauts walk on the Moon, and people perceive the Earth as a fragile whole.

            Budyko and Sellers present models of catastrophic ice-albedo feedbacks.

            The *Nimbus 3* satellite begins to provide comprehensive global atmospheric temperature measurements.

1970   The first Earth Day is celebrated; the environmental movement attains strong influence, spreads concern about global degradation.

            The U.S. National Oceanic and Atmospheric Administration, the world's leading funder of climate research, is created.

            Aerosols from human activity are shown to be increasing swiftly. Bryson claims they are causing global cooling.

1971   An SMIC conference of leading scientists reports a danger of rapid and serious global climate change caused by humans, and calls for an organized research effort.

            The *Mariner 9* spacecraft finds a great dust storm warming the atmosphere of Mars, and indications of a radically different climate in the past.

1972   Ice cores and other evidence show big climate shifts in the past between relatively stable modes in the space of a thousand years or so.

1973   An oil embargo and price rise bring about the first energy
            crisis.

1974   Serious droughts since 1972 increase concern about climate;
            cooling from aerosols is suspected to be as likely as warming;
            journalists talk of a new ice age

1975   Warnings about environmental effects of airplanes lead to an
            investigations of trace gases in the stratosphere and the
            discovery of danger to the ozone layer.

            Manabe and his collaborators produce complex but plausible
            computer models that show a temperature rise of several
            degrees for doubled CO_2.

1976   Studies find that CFCs (1975) and methane and ozone (1976)
            can make a serious contribution to the greenhouse effect.

            Deep-sea cores show a dominating influence from 100,000-year
            Milankovitch orbital changes, which emphasizes the role of
            feedbacks.

            Deforestation and other ecosystem changes are recognized as
            major factors in the future of the climate.

            Eddy shows that there were prolonged periods without sunspots
            in past centuries, which corresponded to cold periods.

1977   Scientific opinion tends to converge on global warming as the
            biggest climate risk in the next century.

1978   Attempts to coordinate climate research in the U.S. end with an
            inadequate National Climate Program Act, accompanied by
            temporary growth in funding.

1979   The second oil energy crisis occurs. A strengthened
            environmental movement encourages renewable energy sources
            and inhibits nuclear energy growth.

            A U.S. National Academy of Sciences report finds it highly
            credible that doubling CO_2 will bring about global warming of
            1.5°–4.5°C.

The World Climate Research Program is launched to coordinate international research.

1981   The election of President Reagan brings a backlash against the environmental movement; political conservatism is linked to skepticism about global warming.

The IBM personal computer is introduced. Advanced economies use increasingly less energy per unit of output.

Hansen and others show that sulfate aerosols can significantly cool the climate, a finding that raises confidence in models showing future greenhouse warming.

Some scientists predict that a greenhouse warming "signal" should be visible by about the year 2000.

1982   Greenland ice cores reveal drastic temperature oscillations in the space of a century in the distant past.

Strong global warming since mid-1970s is reported; 1981 was the warmest year on record.

1983   Reports from the U.S. National Academy of Sciences and Environmental Protection Agency spark conflict; greenhouse warming becomes a prominent topic in mainstream politics.

1985   Ramanathan and his collaborators announce that global warming may come twice as fast as expected, from a rise of methane and other trace greenhouse gases.

The Villach Conference declares a consensus among experts that some global warming seems inevitable; it calls on governments to consider international agreements to restrict emissions.

Antarctic ice cores show that CO_2 and temperature went up and down together through past ice ages, which points to powerful feedbacks.

Broecker speculates that a reorganization of North Atlantic Ocean circulation can bring swift and radical climate change.

1987   The Montreal Protocol of the Vienna Convention requires international restrictions on the emission of ozone-destroying gases.

1988   News media coverage of global warming leaps upward following record heat and droughts as well as testimony by Hansen.

The Toronto Conference calls for strict, specific limits on greenhouse gas emissions; U.K. Prime Minister Thatcher is the first major leader to call for action.

Ice-core and biology studies confirm that living ecosystems make climate feedback by way of methane, which could accelerate global warming.

The Intergovernmental Panel on Climate Change (IPCC) is established.

1989   Fossil-fuel and other U.S. industries form the Global Climate Coalition to tell politicians and the public that climate science is too uncertain to justify action.

1990   The first IPCC report says the world has been warming and future warming seems likely.

1991   Mt. Pinatubo erupts; Hansen predicts a cooling pattern, which will validate (by 1995) computer models of aerosol effects.

Global warming skeptics claim that twentieth-century ' temperature changes followed from solar influences. (The solar-climate correlation would fail in the following decade.)

Studies from 55 million years ago show a possibility that the eruption of methane from the seabed could intesify enormous self-sustained warming.

1992   A conference in Rio de Janeiro produces the U.N. Framework Convention on Climate Change, but the U.S. blocks calls for serious action.

The study of ancient climates reveals climate sensitivity in the same range as that predicted independently by computer models.

1993 Greenland ice cores suggest that great climate changes (at least on a regional scale) can occur in the space of a single decade.

1995 The second IPCC report detects a "signature" of human-caused greenhouse-effect warming; it declares that serious warming is likely in the coming century.

    Reports of the breakup of Antarctic ice shelves and other signs of actual current warming in polar regions begin to affect public opinion.

1997 Toyota introduces the Prius in Japan, the first mass-market gas-electric hybrid car; there is swift progress in large wind turbines and other energy alternatives.

    An international conference produces the Kyoto Protocol, which sets targets to reduce greenhouse gas emissions if enough nations sign onto a treaty.

1998 A "Super El Niño" causes weather disasters and the warmest year on record (approximately matched by 2005 and 2007). Borehole data confirm an extraordinary warming trend.

    Qualms about arbitrariness in computer models diminish as teams model ice-age climate and dispense with special adjustments to reproduce current climate.

2000 The Global Climate Coalition dissolves as many corporations grapple with the threat of warming, but the oil lobby convinces the U.S. administration to deny a problem exists.

    Various kinds of studies emphasize the variability and importance of biological feedbacks in the carbon cycle, which may accelerate warming.

2001 The third IPCC report states baldly that global warming, unprecedented since the end of the last ice age, is "very likely," along with possible severe surprises. Debate effectively ends among all but a few scientists.

    A National Academy panel sees a "paradigm shift" in scientific recognition of the risk of abrupt climate change (decade-scale).

The Bonn meeting, attended by most countries but not the U.S., develops mechanisms for working toward the Kyoto targets.

Warming is observed in ocean basins; the match with computer models gives a clear signature of greenhouse-effect warming.

2003   Numerous observations raise concern that collapse of ice sheets (in West Antarctica and Greenland) can raise sea levels faster than most had believed.

A deadly summer heat wave in Europe accelerates the divergence between European and U.S. public opinion.

2004   In a controversy over temperature data covering the past millennium, most conclude that climate variations were not comparable to post-1980 global warming.

2005   The Kyoto treaty goes into effect, signed by major industrial nations except the U.S. Work to retard emissions accelerates in Japan, in Western Europe, and among U.S. regional governments and corporations.

Hurricane Katrina and other major tropical storms spur debate over the impact of global warming on storm intensity.

2007   The fourth IPCC report warns that serious effects of warming have become evident; the cost of reducing emissions would be far less than the damage they will cause.

The level of CO_2 in the atmosphere reaches 382 ppm. The mean global temperature (a five-year average) is 14.5°C, the warmest in hundreds, perhaps thousands, of years.

For further reading: the literature on current climate science is changing rapidly. For regularly updated suggestions on works in print and on the Internet, visit www.aip.org/history/climate/links.htm.

For complete references and a bibliography of about two thousand items, as well as more extended discussion of all the topics touched on here, visit www.aip.org/history/climate.

1. How Could Climate Change?

1. "Warmer World," *Time*, 2 Jan. 1939, p. 27.
2. Albert Abarbanel and Thomas McCluskey, "Is the World Getting Warmer?" *Saturday Evening Post*, 1 July 1950, p. 63.
3. "Warmer World," p. 27.
4. G. S. Callendar, "The Artificial Production of Carbon Dioxide and Its Influence on Climate," *Quarterly J. Royal Meteorological Society* 64 (1938): 223–240.
5. John Tyndall, "Further Researches on the Absorption and Radiation of Heat by Gaseous Matter" (1862), in Tyndall, *Contributions to Molecular Physics in the Domain of Radiant Heat* (New York: Appleton, 1873), p. 117.
6. John Tyndall, "On Radiation through the Earth's Atmosphere," *Philosophical Magazine* ser. 4, 25 (1863): 204–205.
7. Athelstan Spilhaus, interview by Ron Doel, Nov. 1989, American Institute of Physics, College Park, Md.
8. H. Lamb quoted in Tom Alexander, "Ominous Changes in the World's Weather," *Fortune*, Feb. 1974, p. 90.
9. William Joseph Baxter, *Today's Revolution in Weather* (New York: International Economic Research Bureau, 1953), p. 69.

10. Thomas C. Chamberlin, "On a Possible Reversal of Deep-Sea Circulation and Its Influence on Geologic Climates," *J. Geology* 14 (1906): 371.

11. James R. Fleming, *Historical Perspectives on Climate Change* (New York: Oxford University Press, 1998), chaps. 2–4.

12. Hubert H. Lamb, *Through All the Changing Scenes of Life: A Meteorologist's Tale* (Norfolk, U.K.: Taverner, 1997), pp. 192–193.

2. Discovering a Possibility

1. C.-G. Rossby, "Current Problems in Meteorology," in *The Atmosphere and the Sea in Motion*, ed. Bert Bolin (New York: Rockefeller Institute Press, 1959), p. 15.

2. Gilbert Plass, interview by Weart, 14 March 1996, American Institute of Physics, College Park, Md.

3. G. N. Plass, "Carbon Dioxide and the Climate," *American Scientist* 44 (1956): 302–316.

4. Ibid.

5. Roger Revelle, "The Oceans and the Earth," talk given at American Association for the Advancement of Sciences symposium, 27 Dec. 1955, typescript, folder 66, box 28, Revelle Papers MC6, Scripps Institution of Oceanography archives, La Jolla, Calif.

6. Roger Revelle and Hans E. Suess, "Carbon Dioxide Exchange between Atmosphere and Ocean and the Question of an Increase of Atmospheric CO_2 during the Past Decades," *Tellus* 9 (1957): 18–27.

7. Clark A. Miller, "Scientific Internationalism in American Foreign Policy: The Case of Meteorology, 1947–1958," in *Changing the Atmosphere: Expert Knowledge and Environmental Governance*, ed. Clark A. Miller and Paul N. Edwards (Cambridge, Mass.: MIT Press, 2001), p. 171 and passim.

8. J. A. Eddy, interview by Weart, April 1999, American Institute of Physics, College Park, Md., p. 4.

9. C. C. Wallén, "Aims and Methods in Studies of Climatic Fluctuations," in *Changes of Climate: Proceedings of the Rome Symposium Organized by UNESCO and the World Meteorological Organization, 1961* (UNESCO Arid Zone Research Series, 20) (Paris: UNESCO, 1963), p. 467.

10. Roger Revelle, interview by Earl Droessler, Feb. 1989, American Institute of Physics, College Park, Md.

11. Charles D. Keeling, "The Concentration and Isotopic Abundances of Carbon Dioxide in the Atmosphere," *Tellus* 12 (1960): 200–203.

3. A Delicate System

1. Jhan Robbins and June Robbins, "100 Years of Warmer Weather," *Science Digest,* Feb. 1956, p. 83.

2. Helmut Landsberg, reported in "A Warmer Earth Evident at Poles," *New York Times,* 15 Feb. 1959.

3. United States Congress (85:2), House of Representatives, Committee on Appropriations, *Report on the International Geophysical Year* (Washington, D.C.: Government Printing Office, 1957), p. 104.

4. J. Gordon Cook, *Our Astonishing Atmosphere* (New York: Dial, 1957), p. 121.

5. Conservation Foundation, *Implications of Rising Carbon Dioxide Content of the Atmosphere* (New York: Conservation Foundation, 1963).

6. National Academy of Sciences, Committee on Atmospheric Sciences Panel on Weather and Climate Modification, *Weather and Climate Modification: Problems and Prospects,* 2 vols. (Washington, D.C.: National Academy of Sciences, 1966), vol. 1, p. 10.

7. Ibid., pp. 16, 20.

8. Cesare Emiliani, "Ancient Temperatures," *Scientific American,* Feb. 1958, p. 54.

9. Wallace S. Broecker, "In Defense of the Astronomical Theory of Glaciation," *Meteorological Monographs* 8, no. 30 (1968): 139.

10. Wallace Broecker et al., "Milankovitch Hypothesis Supported by Precise Dating of Coral Reef and Deep-Sea Sediments," *Science* 159 (1968): 300.

11. C. E. P. Brooks, "The Problem of Mild Polar Climates," *Quarterly J. Royal Meteorological Society* 51 (1925): 90–91.

12. David B. Ericson et al., "Late-Pleistocene Climates and Deep-Sea Sediments," *Science* 124 (1956): 388.

13. Wallace Broecker, "Application of Radiocarbon to Oceanography and Climate Chronology" (PhD thesis, Columbia University, 1957), pp. v, 9.

14. Harry Wexler, "Variations in Insolation, General Circulation and Climate," *Tellus* 8 (1956): 480.

15. Wallace Broecker, interview by Weart, Nov. 1997, American Institute of Physics, College Park, Md.

16. Lewis F. Richardson, *Weather Prediction by Numerical Process* (Cambridge: Cambridge University Press, 1922; rpt. New York: Dover, 1965), pp. 219, ix.

17. Jule G. Charney et al., "Numerical Integration of the Barotropic Vorticity Equation," *Tellus* 2 (1950): 245.

18. C.-G. Rossby, "Current Problems in Meteorology," in *The Atmosphere and the Sea in Motion*, ed. Bert Bolin (New York: Rockefeller Institute Press, 1959), p. 30.

19. Norbert Wiener, "Nonlinear Prediction and Dynamics," in *Proceedings of the Third Berkeley Symposium on Mathematical Statistics and Probability*, ed. Jerzy Neyman (Berkeley: University of California Press, 1956), p. 247.

20. Edward N. Lorenz, "Deterministic Nonperiodic Flow," *J. Atmospheric Sciences* 20 (1963): 130, 141.

21. Edward N. Lorenz, "Climatic Determinism," *Meteorological Monographs* 8 (1968): 3.

22. J. Murray Mitchell, "Concluding Remarks" [based on Revelle's summary at the conference], in Mitchell, "Causes of Climatic Change" (*Proceedings*, VII Congress, International Union for Quaternary Research, vol. 5, 1965), *Meteorological Monographs* 8, no. 30 (1968): 157–158.

23. Hubert H. Lamb, "Climatic Fluctuations," in *General Climatology*, ed. H. Flohn (Amsterdam: Elsevier, 1969), p. 178.

4. A Visible Threat

1. Reid A. Bryson and Wayne M. Wendland, "Climatic Effects of Atmospheric Pollution," in *Global Effects of Environmental Pollution*, ed. S. F. Singer (New York: Springer-Verlag, 1970), p. 137.

2. J. Murray Mitchell Jr., "Recent Secular Changes of Global Temperature," *Annals of the New York Academy of Sciences* 95 (1961): 247.

3. SCEP (Study of Critical Environmental Problems), *Man's Impact on the Global Environment: Assessment and Recommendation for Action* (Cambridge, Mass.: MIT Press, 1970), pp. 18, 12.

4. Carroll L. Wilson and William H. Matthews, eds., *Inadvertent Climate Modification: Report of Conference, Study of Man's Impact on Climate (SMIC), Stockholm* (Cambridge, Mass.: MIT Press, 1971), pp. 17, 182, 129.

5. David A. Barreis and Reid A. Bryson, "Climatic Episodes and the Dating of the Mississippian Cultures," *Wisconsin Archeologist* (Dec. 1965): 204.

6. Reid A. Bryson, "A Perspective on Climatic Change," *Science* 184 (1974): 753–760; Reid A. Bryson et al., "The Character of Late-Glacial and Postglacial Climatic Changes (Symposium, 1968)," in *Pleistocene and Recent Environments of the Central Great Plains* (University of Kansas Department of Geology, Special Publication), ed. Wakefield Dort Jr. and J. Knox

Jones Jr. (Lawrence: University Press of Kansas, 1970), p. 72; W. M. Wendland and Reid A. Bryson, "Dating Climatic Episodes of the Holocene," *Quaternary Research* 4 (1974): 9–24.

7. Richard B. Alley, *The Two-Mile Time Machine* (Princeton, N.J.: Princeton University Press, 2000); Paul A. Mayewski and Frank White, *The Ice Chronicles: The Quest to Understand Global Climate Change* (Hanover, N.H.: University Press of New England, 2002).

8. J. Murray Mitchell Jr., "The Natural Breakdown of the Present Interglacial and Its Possible Intervention by Human Activities," *Quaternary Research* 2 (1972): 437–438.

9. W. Dansgaard et al., "Speculations about the Next Glaciation," *Quaternary Research* 2 (1972): 396.

10. Johannes Weertman, "Stability of the Junction of an Ice Sheet and an Ice Shelf," *J. Glaciology* 13 (1974): 3.

11. George J. Kukla and R. K. Matthews, "When Will the Present Interglacial End?" *Science* 178 (1972): 190–191.

12. Christian E. Junge, "Atmospheric Chemistry," *Advances in Geophysics* 4 (1958): 95.

13. J. Murray Mitchell Jr., "A Preliminary Evaluation of Atmospheric Pollution as a Cause of the Global Temperature Fluctuation of the Past Century," in *Global Effects of Environmental Pollution,* ed. S. Fred Singer (New York: Springer-Verlag, 1970), p. 153.

14. S. Ichtiaque Rasool and Stephen H. Schneider, "Atmospheric Carbon Dioxide and Aerosols: Effects of Large Increases on Global Climate," *Science* 173 (1971): 138.

15. G. D. Robinson, "Review of Climate Models," in *Man's Impact on the Climate* (Study of Critical Environmental Problems [SCEP] Report), ed. William H. Matthews et al. (Cambridge, Mass.: MIT Press, 1971), p. 214.

16. Mikhail I. Budyko, "The Effect of Solar Radiation Variations on the Climate of the Earth," *Tellus* 21 (1969): 618.

17. William D. Sellers, "A Global Climatic Model Based on the Energy Balance of the Earth-Atmosphere System," *J. Applied Meteorology* 8 (1969): 392.

18. Owen B. Toon et al., "Climatic Change on Mars and Earth," in *Proceedings of the WMO/IAMAP Symposium on Long-Term Climatic Fluctuations, Norwich, Aug. 1975* (WMO Doc. 421) (Geneva: World Meteorological Organization, 1975), p. 495.

5. Public Warnings

1. William A. Reiners, "Terrestrial Detritus and the Carbon Cycle," in *Carbon and the Biosphere,* ed. George M. Woodwell and Erene V. Pecan (Washington, D.C.: Atomic Energy Commission [National Technical Information Service, CONF-7502510], 1973), p. 327.

2. Tom Alexander, "Ominous Changes in the World's Weather," *Fortune,* Feb. 1974, p. 92.

3. "Another Ice Age?" *Time,* 26 June 1974, p. 86.

4. G. S. Benton quoted in H. M. Schmeck Jr., "Scientist Sees Man's Activities Ruling Climate by 2000," *New York Times,* 30 April 1970.

5. Lowell Ponte, *The Cooling* (Englewood Cliffs, N.J.: Prentice-Hall, 1976), pp. 234–235.

6. "The Weather Machine," BBC-television and WNET, expanded in a book: Nigel Calder, *The Weather Machine* (New York: Viking, 1975), quote on p. 134.

7. John Gribbin, "Man's Influence Not Yet Felt by Climate," *Nature* 264 (1976): 608; B. J. Mason, "Has the Weather Gone Mad?" *New Republic,* 30 July 1977, pp. 21–23.

8. Stephen H. Schneider with Lynne E. Mesirow, *The Genesis Strategy: Climate and Global Survival* (New York: Plenum Press, 1976), esp. chap. 3.

9. Gerald Stanhill, "Climate Change Science Is Now Big Science," *Eos, Transactions of the American Geophysical Union* 80, no. 35 (1999): 396 (from graph).

10. Joseph Smagorinsky, "Numerical Simulation of the Global Circulation," in *Global Circulation of the Atmosphere,* ed. G. A. Corby (London: Royal Meteorological Society, 1970), p. 33.

11. Wallace Broecker, interview by Weart, Nov. 1997, American Institute of Physics, College Park, Md.

12. P. H. Abelson, "Energy and Climate," *Science* 197 (1977): 941.

13. "The World's Climate Is Getting Worse," *Business Week,* 2 Aug. 1976, p. 49; "CO_2 Pollution May Change the Fuel Mix," *Business Week,* 8 Aug. 1977, p. 25.

14. National Academy of Sciences, Climate Research Board, *Carbon Dioxide and Climate: A Scientific Assessment* (Washington, D.C.: National Academy of Sciences, 1979), pp. 2, 3 (the Charney Report); Nicholas Wade, "CO_2 in Climate: Gloomsday Predictions Have No Fault," *Science* 206 (1979): 912–913.

15. National Academy of Sciences, Committee on Atmospheric Sciences, Panel on Weather and Climate Modification, *Weather and Climate Modification: Problems and Prospects,* 2 vols. (Washington, D.C.: National Academy of Sciences, 1966), vol. 1, p. 11.

16. E.g., E. P. Stebbing, "The Encroaching Sahara: The Threat to the West African Colonies," *Geographical J.* 85 (1935): 523.

17. Charles D. Keeling, "The Carbon Dioxide Cycle: Reservoir Models to Depict the Exchange of Atmospheric Carbon Dioxide with the Ocean and Land Plants," in *Chemistry of the Lower Atmosphere,* ed. S. I. Rasool (New York: Plenum Press, 1973), p. 320.

18. Ibid., p. 279.

19. George M. Woodwell, "The Carbon Dioxide Question," *Scientific American,* Jan. 1978, p. 43.

20. Wallace S. Broecker et al., "Fate of Fossil Fuel Carbon Dioxide and the Global Carbon Budget," *Science* 206 (1979): 409, 417.

21. Opinion Research Corporation polls, May 1981, USORC.81MAY.R22, and April 1980, USORC.80APR1.R3M. Data furnished by Roper Center for Public Opinion Research, Storrs, Conn.

6. The Erratic Beast

1. Ed Lorenz, address to the American Association for the Advancement of Science, Washington, D.C., 29 Dec. 1979.

2. James E. Hansen et al., "Climate Impact of Increasing Atmospheric Carbon Dioxide," *Science* 213 (1981): 961.

3. National Academy of Sciences, Climate Research Board, *Carbon Dioxide and Climate: A Scientific Assessment* (Washington, D.C.: National Academy of Sciences, 1979), p. 2.

4. Hansen, "Climate Impact," p. 957; Roland A. Madden and V. Ramanathan, "Detecting Climate Change Due to Increasing Carbon Dioxide," *Science* 209 (1980): 763–768.

5. Stephen H. Schneider, "Introduction to Climate Modeling," in *Climate System Modeling,* ed. Kevin E. Trenberth (Cambridge: Cambridge University Press, 1992), p. 26.

6. Wallace S. Broecker, "Climatic Change: Are We on the Brink of a Pronounced Global Warming?" *Science* 189 (1975): 460–464.

7. Hans E. Suess, "Climatic Changes, Solar Activity, and the Cosmic-Ray Pro-

duction Rate of Natural Radiocarbon," *Meteorological Monographs* 8, no. 30 (1968): 146.

8. R. E. Dickinson, "Solar Variability and the Lower Atmosphere," *Bulletin of the American Meteorological Society* 56 (1975): 1240–1248.

9. Eddy, interview by Weart, April 1999, American Institute of Physics, College Park, Md.

10. Raymond S. Bradley, *Quaternary Paleoclimatology: Methods of Paleoclimatic Reconstruction* (Boston: Allen and Unwin, 1985), p. 69.

11. J.-R. Petit quoted in Gabrielle Walker, "The Ice Man [Interview with Jean-Robert Petit]," *New Scientist,* 29 Jan. 2000, pp. 40–43.

12. National Academy of Sciences, United States Committee for the Global Atmospheric Research Program (GARP), *Understanding Climatic Change: A Program for Action* (Washington, D.C.: National Academy of Sciences, 1975), p. 4.

13. Kirk Bryan, "Climate and the Ocean Circulation. III. The Ocean Model," *Monthly Weather Review* 97 (1969): 822.

14. R. O. Reid et al., *Numerical Models of World Ocean Circulation* (Washington, D.C.: National Academy of Sciences, 1975), p. 3.

15. James E. Hansen et al., "Climate Response Times: Dependence on Climate Sensitivity and Ocean Mixing," *Science* 229 (1985): 857–859.

16. W. Dansgaard et al., "A New Greenland Deep Ice Core," *Science* 218 (1982): 1273.

17. U. Siegenthaler et al., "Lake Sediments as Continental Delta O^{18} Records from the Glacial/Post-Glacial Transition," *Annals of Glaciology* 5 (1984): 149.

18. Broecker, "The Biggest Chill," *Natural History,* Oct. 1987, pp. 74–82; Broecker et al., "Does the Ocean-Atmosphere System Have More Than One Stable Mode of Operation?" *Nature* 315 (1985): 21–25.

19. Broecker, "The Biggest Chill," p. 82.

7. Breaking into Politics

1. Albert Gore Jr., *Earth in the Balance: Ecology and the Human Spirit* (Boston: Houghton Mifflin, 1992), pp. 4–6.

2. Robert G. Fleagle, "The U.S. Government Response to Global Change: Analysis and Appraisal," *Climatic Change* 20 (1992): 72.

3. James E. Jensen, "An Unholy Trinity: Science, Politics and the Press" (unpublished talk), 1990.

4. Walter Sullivan, "Study Finds Warming Trend That Could Raise Sea Level,"

New York Times, 22 Aug. 1981, p. 1, and "Heating Up the Atmosphere," 29 Aug. 1981, p. 22.

5. National Academy of Sciences, Carbon Dioxide Assessment Committee, *Changing Climate* (Washington, D.C.: National Academy of Sciences, 1983), p. 3.

6. Stephen Seidel and Dale Keyes, *Can We Delay a Greenhouse Warming?* 2nd ed. (Washington, D.C.: Environmental Protection Agency, 1983), pp. ix, 7 (of sect. 7); *New York Times,* 18 Oct. 1983, p. 1.

7. Bert Bolin et al., eds., *The Greenhouse Effect, Climatic Change, and Ecosystems* (SCOPE Report No. 29) (Chichester and New York: John Wiley, 1986), pp. xx–xxi.

8. Jonathan Weiner, *The Next One Hundred Years: Shaping the Fate of Our Living Earth* (New York: Bantam, 1990), p. 79.

9. Stephen H. Schneider, "An International Program on 'Global Change': Can It Endure?" *Climatic Change* 10 (1987): 215.

10. Michael McElroy quoted in Andrew C. Revkin, "Endless Summer: Living with the Greenhouse Effect," *Discover,* Oct. 1988, p. 61.

11. Philip Shabecoff, "Global Warming Has Begun, Expert Tells Senate," *New York Times,* 24 June 1988, p. 1.

12. Spencer R. Weart, *Never at War: Why Democracies Will Not Fight One Another* (New Haven: Yale University Press, 1998), p. 265.

8. Speaking Science to Power

1. Published surveys include Stanley A. Chagnon et al., "Shifts in Perception of Climate Change: A Delphi Experiment Revisited," *Bulletin of the American Meteorological Society* 73, no. 10 (1992): 1623–1627, and David H. Slade, "A Survey of Informal Opinion Regarding the Nature and Reality of a 'Global Greenhouse Warming,'" *Climatic Change* 16 (1990): 1–4.

2. Frederick Seitz, ed., *Global Warming Update: Recent Scientific Findings* (Washington, D.C.: George C. Marshall Institute, 1992), p. 28.

3. L. Roberts, "Global Warming: Blaming the Sun," *Science* 246 (1989): 992–993.

4. Robert Lichter, "A Study of National Media Coverage of Global Climate Change, 1985–1991" (Washington, D.C.: Center for Science, Technology and Media, 1992).

5. Philip Shabecoff, "Bush Denies Putting Off Action on Averting Global Climate Shift," *New York Times,* 19 April 1990, p. B4.

6. Tom M. L. Wigley, "Outlook Becoming Hazier," *Nature* 369 (1994): 709–710.

7. This preliminary version was quoted in the press; the final phrasing was: "The observed warming trend is unlikely to be completely natural in origin." Intergovernmental Panel on Climate Change, *Climate Change 1995: The Science of Climate Change,* ed. J. T. Houghton et al. (Cambridge: Cambridge University Press, 1996), p. 22. This and other IPCC reports cited in notes below may be found at www.ipcc.ch.

8. Richard A. Kerr, "It's Official: First Glimmer of Greenhouse Warming Seen," *Science* 270 (1995): 1565–1567.

9. T. M. L. Wigley and P. M. Kelly, "Holocene Climatic Change, ^{14}C Wiggles and Variations in Solar Irradiance," *Philosophical Transactions of the Royal Society of London* A330 (1990): 558.

10. Intergovernmental Panel, *Summary for Policymakers. The Regional Impacts of Climate Change: An Assessment of Vulnerability. A Special Report of IPCC Working Group II,* ed. R. T. Watson et al. (Cambridge: Cambridge University Press, 1997), p. 6.

11. Wallace S. Broecker, "Thermohaline Circulation, the Achilles Heel of Our Climate System: Will Man-Made CO_2 Upset the Current Balance?" *Science* 278 (1997): 1582–1588.

12. National Academy of Sciences, Committee on Abrupt Climate Change, *Abrupt Climate Change: Inevitable Surprises* (Washington, D.C.: National Academy of Sciences, 2002), pp. 16, 121; Intergovernmental Panel, *Climate Change 1995,* p. 7.

9. The Work Completed . . . and Begun

1. Philip Shabecoff, "Draft Report on Global Warming Foresees Environmental Havoc in U.S.," *New York Times,* 20 Oct. 1988, reporting on draft of U.S. Environmental Protection Agency, *The Potential Effects of Global Climate Change on the United States* (EPA-230-5-89-050) (Washington, D.C.: Environmental Protection Agency, 1989).

2. Hans Joachim Schellnhuber, quoted in Walter Gibbs and Sarah Lyall, "Gore Shares Peace Prize for Climate Change Work," *New York Times,* 13 Oct. 2007.

3. Intergovernmental Panel on Climate Change, *Climate Change 2001: The Scientific Basis. Contribution of Working Group I to the Third Assessment Re-*

port of the IPCC, ed. J. T. Houghton et al. (Cambridge: Cambridge University Press, 2001).

4. John Browne, speech at Stanford University, 19 May 1997, at www.gsb.stanford.edu/community/bmag/sbsm0997/feature_ranks.html.

5. Jaclyn Marisa Dispensa and Robert J. Brulle, "Media's Social Construction of Environmental Issues: Focus on Global Warming—A Comparative Study," *International J. of Sociology and Social Policy* 23 (2003): 74.

6. Sydney Levitus et al., "Anthropogenic Warming of Earth's Climate System," *Science* 292 (2001): 267–270.

7. The best series of polls are from Gallup (subscription required) and the Pew Research Center for the People & the Press. Many other national and international polls may be found by searching the Internet.

8. John Immerwahr, *Waiting for a Signal: Public Attitudes toward Global Warming, the Environment and Geophysical Research* (New York: Public Agenda, 1999), online at http://Earth.agu.org/sci_soc/sci_soc.html; summary in Randy Showstock, "Report Suggests Some Public Attitudes about Geophysical and Environmental Issues," *Eos, Transactions of the American Geophysical Union* 80, no. 24 (1999): 269, 276.

9. Wallace S. Broecker, "Future Global Warming Scenarios" (letter), *Science* 304 (2004): 388.

10. *Time*, 3 Oct. 2006, front cover; Andrew Revkin, "Meltdown," *New York Times*, 23 April 2006 (Week in Review).

11. "Global Warming," *Business Week*, 18 Dec. 2006, p. 102.

12. Jeffrey Ball, "New Consensus: In Climate Controversy, Industry Cedes Ground," *Wall Street Journal*, 23 Jan. 2007, p. 1.

13. Sergei Kirpotin quoted in Fred Pearce, "Climate Warming as Siberia Melts," *New Scientist*, 13 Aug. 2005, p. 12.

14. Robert Bindschadler quoted in Larry Rohter, "Antarctica, Warming, Looks More Vulnerable," *New York Times*, 25 Jan. 2005, sec. D.

15. Intergovernmental Panel, *Climate Change 2007: The Physical Basis of Climate Change. Contribution of Working Group I to the Fourth Assessment Report of the IPCC*, ed. Susan Solomon et al. (Cambridge: Cambridge University Press, 2007), pp. 10, 13.

16. A. Barrie Pittock, "Are Scientists Underestimating Climate Change?" *Eos, Transactions of the American Geophysical Union* 87 (2006): 340.

17. Intergovernmental Panel, "Draft Summary for Policymakers," ed. Lenny Bernstein et al., pp. 1–23 in *Climate Change 2007: Synthesis Report of the In-*

tergovernmental Panel on Climate Change Fourth Assessment Report (Cambridge: Cambridge University Press 2007), pp. 12–13.

18. Nicholas Stern, *The Economics of Climate Change: The Stern Review* (Cambridge: Cambridge University Press and HM Treasury, 2006), p. 3; online at   www.hm-treasury.gov.uk/independent_reviews/stern_review_economics _climate_change/stern_review_report.cfm.

Reflections

1. The bibliography is at www.aip.org/history/climate/bib.htm.
2. Intergovernmental Panel on Climate Change, *Climate Change 2007: The Physical Basis of Climate Change. Contribution of Working Group I to the Fourth Assessment Report of the IPCC,* ed. Susan Solomon et al. (Cambridge: Cambridge University Press, 2007).